P9-DVU-129

An Introduction to Crop Physiology

An Introduction to Crop Physiology

F. L. MILTHORPE AND J. MOORBY

School of Biological Sciences, Macquarie University

CAMBRIDGE UNIVERSITY PRESS

Published by the Syndics of the Cambridge University Press
Bentley House, 200 Euston Road, London NW1 2DB
American Branch: 32 East 57th Street, New York, N.Y.10022

Library of Congress Catalogue Card Number: 73–80468

ISBNS:
0 521 20260 4 hard covers
0 521 09816 5 paperback

First published 1974

Printed in Great Britain
at the University Printing House, Cambridge
(Brooke Crutchley, University Printer)

Contents

Preface

The early physiologists, in setting out to understand how a plant works, soon discovered sets of phenomena which could be grouped as 'processes'. Further study has generally been in the direction of analysing these processes in greater depth. This reduction approach, with the wealth of resulting detail, has led to the corpus of knowledge being divided into groupings such as molecular biology, cell biology, biochemistry, biophysics, and process physiology (usually called plant physiology). Concurrently, exploration of the behaviour of communities of plants and their dependent associated organisms in their natural environments led to the development of ecology. We are interested in the hinterland between ecology and process physiology and are essentially concerned with understanding the ways in which the various processes are integrated to produce the response shown by whole plants when they are grown as a community in a natural environment. The subject matter covered has close affinities with that variously known as whole-plant physiology, environmental plant physiology, growth physiology, autecology and agronomy, although each of these terms means different things to different people. Most of the material we deal with concerns agricultural systems because they are so important and because they have been the subject of the most intensive study. Many of the concepts and approaches, however, are applicable to and invaluable in understanding the more complex natural ecosystems. We use the term 'physiology' rather than 'ecology' because we ignore all interrelationships with other organisms except higher plants. Emphasis is placed on integration and on trying to assess quantitatively the importance of each component in the context of the whole plant–environment system. Because of its importance in agricultural systems our major concern is the understanding and prediction of productivity, i.e. of putting plant physiology to work.

We expect that those well indoctrinated in the traditional approach of delving deeper and deeper into the understanding of mechanisms might regard this concern with rate modulation in the organism as a whole as being rather superficial. We would remind such readers of the continued necessity to build back understanding into the wider system whilst recognizing the inherent dangers of superficial extrapolation over a range as wide as from the functioning of cells to the responses of an organism to its environment.

This book is written as an introductory text. As such it aims to present an approach to the subject in a coherent and consecutive manner and with a modest degree of substantiation. We assume that those using it will have already taken an introductory course in plant (i.e. process) physiology and are therefore reasonably familiar with the general features of the functioning of a plant. It would also be helpful to have a good background in soil science and micrometeorology but, for those who have not, we have provided a potted summary of the essential features. Similarly, we have tried to provide enough basic physiology to allow non-physiologists to follow the train of argument.

The most difficult issue we faced was, not surprisingly, what to leave out. To augment coherence as well as to reduce sheer volume, we have kept to the most common types of situation. We also think it essential that actual magnitudes be appreciated. It is not enough to know, for example, that photosynthesis occurs; we need to appreciate the actual rates and the degree of variation in different circumstances. We have tried to meet this need but we warn the reader against too hasty or too wide an extrapolation. We regret that the issue of size has prevented us from acknowledging ideas and concepts where we would have liked to have done so. The reader should however be able to trace the wider literature through the key references appended. A detailed account of the physiology of specific crops is given in *Crop physiology – some case histories*, edited by L. T. Evans and published by Cambridge University Press; this we regard as a complementary volume explaining how the principles explained here have been applied in specific situations.

We owe a very large debt to our fellow workers in this field; their many contributions have made this book possible and we regret that we have not been able to do full justice to, or to acknowledge, their work adequately. Our immediate colleagues have generously read drafts of the text and offered many useful suggestions. We owe special thanks to Dr R. M. Gifford for so readily providing us with the data for Table 5.2. The tremendous labour of getting diagrams into a reproducible form was borne by Miss Betty Thorn and the task of translating our hieroglyphics into legibility by Miss Joyce Chatfield.

We are most grateful to the many authors whose work we have used as illustrations, and to their publishers for so readily allowing us to reproduce their work. These are acknowledged in the text where appropriate.

F.L.M.

1973 J.M.

Acknowledgements

Permission for reproduction was granted by Academic Press and editors of *Advances in Agronomy* (Fig. 7.7; Table 3.2); editors of *American Journal of Botany* (Fig. 6.4); American Society of Agronomy and editors of *Agronomy Journal* and *Crop Science* (Figs. 3.8, 3.9, 5.7, 5.16, 5.24, 9.16; Tables 6.1, 6.4); American Society of Plant Physiologists and editors of *Plant Physiology* (Figs. 3.3, 9.9); Annual Reviews Inc. (Fig. 9.6); Australian Institute of Agricultural Science (Table 9.2); Basic Books Inc. (Fig. 9.5); Blackwell Scientific Publications and editors of *Journal of Applied Ecology* and *New Phytologist* (Figs. 2.10, 2.11, 2.12, 3.1, 3.7, 4.3, 4.4, 5.25, 6.5; Tables 4.6, 4.8); Centre for Agricultural Publishing and Documentation and editors of *European Potato Journal* (Figs. 3.4, 5.28, 6.3); Clarendon Press and editors of *Annals of Botany*, *Journal of Experimental Botany* and *Journal of Soil Science* (Figs. 2.1, 4.2, 4.8, 4.9, 5.17, 5.18, 6.1, 6.2, 7.1, 7.2, 9.4, 9.5, 9.10; Tables 4.2, 5.3); Commonwealth Scientific and Industrial Research Organization and editors of *Australian Journal of Agricultural Research*, *Australian Journal of Biological Sciences* (Figs. 3.3, 4.10, 5.6, 5.10, 5.13, 5.15, 5.19, 5.29, 7.1, 7.3, 7.6, 8.1, 8.3, 8.4, 8.6, 9.2, 9.7, 9.12, 9.14; Tables 4.1, 5.4); Company of Biologists (Figs. 2.4, 9.13); editors of *El Riso* (Fig. 9.3); Elsevier Publishing Co. and editors of *Agricultural Meteorology* (Figs. 5.2, 5.22); Food and Agricultural Organization (Table 9.1); Imperial Chemical Industries and editors of *Outlook on Agriculture* (Fig. 4.1); editors of *Journal of Agricultural Science* (Fig. 9.8; Table 4.7); Professor E. W. Russell and Longman Ltd (Fig. 2.6); Macmillan Ltd and editors of *Nature* (Figs. 5.31, 9.1); Martinus Nijhoff and editors of *Plant and Soil* (Figs. 4.5, 4.7); National Research Council of Canada and editors of *Canadian Journal of Botany* (Figs. 3.3, 7.4); editors of *Netherlands Journal of Agricultural Science* (Fig. 5.1); editors of *Photosynthetica* (Figs. 5.8, 5.9, 5.11, 5.14, 5.27); editors of *Physiologia Plantarum* (Fig. 5.12); Royal Meteorological Society (Fig. 3.5); Springer-Verlag and editors of *Planta* (Figs. 5.4, 5.5, 5.20, 5.21); UNESCO (Fig. 9.14); University of Chicago Press and editors of *American Naturalist* (Fig. 7.8*a*); University of Tokyo (Fig. 5.26); editors, Department of Crop Science, University of West Indies (Fig. 9.11); Veenman and Zonen (Fig. 5.30); Williams and Wilkins Co. and editors of *Soil Science* (Table 4.3); and World

Meteorological Organization (Fig. 5.23). The following authors as owners of copyright or of unpublished data graciously gave permission: Mr E. B. Boerema (Fig. 9.3), Professor A. H. Bunting and Dr D. S. H. Drennan (Figs. 7.4, 7.5); Mr E. K. Christie (Table 4.5); Dr D. J. C. Friend (Fig. 7.5); Dr R. M. Gifford (Table 5.2); Dr H. W. Howard (Table 9.3); Dr H. Idris (Fig. 8.5); Dr J. W. Patrick (Fig. 8.2); Dr D. W. Puckridge (Fig. 7.4); Drs R. S. Russell and P. Newbould (Table 4.4; Fig. 4.11); Dr G. N. Thorne (Table 8.1) and Dr T. E. Williams (Fig. 7.8*b*).

1

Prolegomena

AN OUTLINE OF THE SYSTEM

A crop is an aggregation of individual plants, usually of the same species and normally of very similar genetic constitution (i.e. of the same cultivar), grown in a particular location for a specific product required by man. Conceptually, we may regard the system as consisting of a number of components: the size, number and nature of the initial propagules, the functioning and growth of the plants during their ontogeny, and the environment in which these events occur – an environment imposed by outside influences and modified by the size, form and arrangement of the constituent plants. The interplay of these components and their resultant – the yield and quality of the product sought by man – is our main concern.

It is convenient to take as the central theme the functioning of the plants during their ontogeny from planting to harvest. However, we must recognize that the seeds or vegetative propagules used by the farmer are already fully differentiated individuals, each with a determined genetic complement, possessing one or more central axes, and provided with a supply of food materials. After emergence, the plant becomes fully autotrophic, manufacturing its own constituents from carbon dioxide, water and mineral elements and transforming radiant energy into a usable chemical form. The crop can then be regarded as a group of individual self-expanding factories in which some of the products are re-invested in making new machinery, some are used for making the structure (or fabric) and some go into storage. The net daily production depends, therefore, on the number of machines and the rate at which each works, thereby at any given time reflecting the sizes of the plants and the rates of activity of many physiological processes; the latter in turn, as well as the relative proportions of the daily product diverted towards the different ends, are influenced by the environment and by a series of internal control mechanisms.

Interrelationships between physiological activity and the environment exist throughout ontogeny. Although differing in degree, a great deal is common. It is therefore appropriate to examine this common territory in some detail before proceeding to the major theme; in effect, to explore at one point of time the main features of the major supply functions of the plant. These are essentially photosynthesis and the absorption of mineral nutrients and the transport of their products within the plant. We need also to draw attention to those features of the environment itself which are immediately relevant; what we provide is no more than the barest background, and the reader is advised to delve more deeply in the standard texts of soil science, micrometeorology and plant physiology.

One set of environmental–physiological interrelationships – that concerning water – is peculiar; peculiar in the sense that these interrelationships greatly influence the milieu in which the physiological processes proceed but not through supply for direct participation in metabolic processes. It is true that water is an essential substrate in photosynthesis and hydrolytic reactions, but long before it becomes sufficiently short to influence these directly, effects arising from lack of turgor have completely disrupted the whole system. Water is therefore sufficiently removed functionally from the main theme and so dominant a component of the environment that it deserves a prefatory chapter. The other dominant component on regional and seasonal scales, temperature, does not receive such singular attention; it is also involved indirectly by influencing rates of chemical processes, and through enzyme inactivation, but these can be best dealt with as we proceed.

Variation

All species have a great deal in common, so much so that it is convenient to think in terms of a *generalized plant* and to deal with the differences which occur as *variations*. Where these differences between plants remain relatively constant from generation to generation and between different environments, we think of them as *genetic variations*. We will attempt as we proceed to indicate the significance of some differences between different species of crop plants but, because of our limited objectives, readers will need to refer to more detailed texts for a full discussion. Here we will be content to assess the immediate significance of such variations and to account for their effect in a purely empirical manner. A more fundamental approach would be to start with the gene complement and follow through their effects as expressed in the changing state of the plant and as modified by environment; such is, alas, far beyond our capacity.

Where the variations in any one species grown in different environments are appreciable, they may be regarded as *environmental variations*. There is also another source of variation: that which arises as the plant proceeds in its ontogeny and which may be regarded as changes in its internal state. Such *ontogenetic changes* may be so dramatic – as when the apex changes from a vegetative to a flowering state – that they almost certainly represent the switching-off of one set of genes and bringing another into play. Other changes are more gentle, reflecting a gradual drift with time. Here there may also be some influences arising through differences in the type of messenger-RNA produced; equally there are effects arising from different rates of supply of substrates and their flow to and use in sinks varying in number and intensity. Whatever the source, the ontogenetic drifts are usually large and must always be kept in mind. One consequence is that the degree of response of a plant to a given environmental factor varies greatly during ontogeny. Another is an increasing degree of variation between the members of a population with the same genetic constitution grown in the same environment. Minute differences in initial size, and small transient differences in environment or physiological intensity, lead to differences which become more and more exaggerated as the plants grow. There are still some experimenters who blithely assume that if they work with a genetic clone all difficulties associated with variation will disappear; in fact, because of the sheer impracticability of obtaining clonal propagules of the same size and state, the variation is usually greater than in a population grown from seed of a narrow size range.

THE SIGNIFICANCE OF MERISTEMS

Plant growth is expressed by the division of cells in localized regions and by the expansion and differentiation of these cells. Although variation in the latter can give rise to perceptible differences in the size of an organ, almost all of the differences in size and general form of an organ, as well as the organism, arise from differences in number, activity and persistence of the centres of cell division. The meristems located at the root and shoot apices of the embryo recommence activity following sowing of the seed and persist for most of the life of the plant. That of the stem, which often ceases activity only as a consequence of flower formation, produces the cells forming the basic framework of the primary stem axis. Cell division also recommences in the leaf primordia, and in the cotyledons of some species, and bud primordia are produced in the axils of the leaf primordia. The leaf meristem divides continuously from the time of initiation but the duration of activity is restricted, resulting in the leaf being an organ of limited growth. That in the axillary bud usually continues for a short period after the bud is formed, producing new leaf and bud primordia, and then ceases. Activity may or may not be resumed; if it is, the bud grows out as a branch, or secondary stem axis. The development of the cambium provides another meristem which is persistent throughout the life of the stem. Several orders of lateral roots may be formed, again with persistent terminal and cambial meristems. Leaves, flowers and fruits appear to be the only organs with meristems of limited duration.

Whereas leaf primordia arise from the outer layers of cells in the stem, flower primordia involve all layers and ultimately lead to the cessation of growth of the stem axes on which they are borne. This can have important consequences for the survival of the individual plant. If most of the axes flower at the same time, then the whole plant dies soon after. If this occurs in the first season of growth, the plants are called *annuals*, if in the second season, *biennials*. True *perennials* are plants which always maintain a number of vegetative stem apices among those which are flowering. Some species, such as the century plants, may take many years before they flower; when they do the entire axis (and plant) dies. In others, such as the tomato, when the terminal meristem forms an inflorescence the next lowest bud grows out as a vegetative shoot. Although it also soon flowers, the cycle is repeated continuously.

As most of the differences in size, at least to a first approximation, can be traced to differences in cell number, meristematic activity is an important process which we will consider later. Here, it may be worth reproducing some vital statistics of an average cell:

Mature cortical cell

Volume: 2×10^{-7} cm^3 ($\equiv$ cylinder 100 μm long and 50 μm in diameter; variation in length 50–300 μm and breadth 30–100 μm).

Dry weight: 6×10^{-12} kg (consisting, say, of $\frac{1}{3}$ cell wall (cellulose, pectins, etc), $\frac{1}{3}$ protein and $\frac{1}{3}$ other substances).

Meristematic cell

About $\frac{1}{20}$ of volume and $\frac{1}{10}$ of dry weight of mature cell.

GROWTH, DIFFERENTIATION AND CO-ORDINATION

The development of a plant usually proceeds in a most orderly sequence of changes, leading to an irreversible increase in size and a predictable alteration of form. Size may be measured as height, volume, fresh weight, dry weight, etc. Dry weight is usually preferred as it avoids complications arising from short-term fluctuations in water content and also gives a good indication of the quantity of energy fixed. (Most dried plant tissue, excluding seeds of high protein or fat content, has an energy content of about 1.74×10^7 J kg^{-1}.)

In a general sense, the differentiation of cells, tissues and organs appears to be under fairly strict internal control. The changes in form in each of the species of crop plants, over the range of environments in which they are grown, are usually small; the main differences are in size and in the time taken to reach a given size, i.e. the rate of growth. However, there are notable exceptions, as with flowering, branching, formation of vegetative storage organs such as tubers, and the relative proportions of different substances in seeds and fruits leading to differences in quality. This control and the switching mechanisms, such as those involved with leaf initiation, are still little understood and will generally remain outside this discussion. In these and in more variable responses such as flowering and branching, growth substances are believed to play an important role. Again, we will give less attention to these aspects than some may think warranted. This decision was taken because so little is known about the flow, concentration and responses to these substances in the normal functioning plant and because our main theme concerns the integrated whole plant and its variation with time and environment. Much of the behaviour can be explained directly in terms of supply functions of energy, carbon dioxide, water and mineral elements and the environmental control of rates of supply and growth, and some attempt can be made to put these together in a co-ordinated and quantitative way. Such is not possible with growth-substance – and genetic – effects and so these have to be treated empirically. Hence, a great deal of speculative treatment has been omitted – not because it is unimportant but because it is irrelevant to our immediate objective and is covered adequately elsewhere.

TOWARDS THE QUANTITATIVE

We can usually recognize a number of stages in the long, tortuous and often chaotic process by which man gropes for an understanding of phenomena. An early stage is the recognition that certain events occur and are associated with each other, for example, that most of the dry matter of the plant arises from the transmutation in light of carbon dioxide absorbed by the leaves and water absorbed from the soil. A second stage consists of exploring the events concerned in this process and of obtaining a full, detailed and consistent understanding. Most of the study of biology at present is concerned with this type of activity. Having obtained sufficient qualitative appreciation of the various components of the system, it is then possible to measure them and relate each to the other. This involves finding suitable relationships between the magnitude of a process and the variables which determine it, of putting these together in such a way that the magnitude in any stated set of circumstances can be predicted, and of then assessing the interplay between the different component processes so that the behaviour of the whole can be predicted. It is only when this stage is reached that a science can be said to be truly quantitative; it is not enough to be merely numerical or to assess the degree of probability that two or more events differ from each other. Although we recognize that there still remain large gaps in our qualitative appreciation, especially in respect of growth substances, we believe we should move towards this third stage wherever and to the extent that we can. To this end, numerical functions will be used wherever possible. We appreciate that these will not always be developed fully and rigorously and that we may sometimes assume a background knowledge different from that of the reader. It should be possible, however, to make good any deficit using the references provided. Because of the wide range of topics covered, some letters have been used as symbols for different entities. We have tried to avoid any confusion by the use of appropriate subscripts. A list of all the symbols used is provided (pp. 188–93).

2

The Environment

THE SOIL ENVIRONMENT

Soil structure

We may regard the soil as a matrix which supports the plants and provides supplies of water and minerals. It consists of mineral particles, organic matter, water, air, and living organisms. The mineral particles are derived ultimately from the disintegration and decomposition of igneous rocks but may have been through one or more cycles of erosion and deposition. They consist of unaltered fragments of quartz, felspars and other minerals which make up the sand (2–0.02 mm diameter) and silt (0.02–0.002 mm diameter) fractions; secondary minerals formed during the weathering of rocks which comprise the clay fraction (< 0.002 mm diameter); and compounds such as calcium carbonate, calcium phosphate, and hydrated oxides of iron and other elements. The clay minerals are layer-lattice minerals in which replacement of silicon atoms by aluminium or aluminium by magnesium leads to the whole crystal lattice having a net negative charge. These charges are neutralized by adsorbed cations which can easily exchange with other cations and thus provide a source readily available to the plant. The organic matter consists of a whole range of materials from undecayed plant and animal remains through many transient forms to the fairly stable, dark-coloured, amorphous humus.

The mineral matter (90–98 per cent) and organic matter (2–10 per cent) are arranged to give a definite structure to the soil – although probably few soil scientists would agree on the finer details of this structure. We will take as a working concept a picture due mainly to Emerson (1959). This visualizes micro-aggregates consisting of quartz (and felspar) particles being held together by organic matter and groups of clay crystals somewhat in the way shown in Fig. 2.1. The proportions of the three components vary, of course, between different soils. The micro-aggregates are relatively transient, being broken down and reforming under the influence of tillage implements, rain, micro-organisms, etc., and are joined together into larger aggregates or crumbs, which in a well-structured soil are 1–5 mm in diameter. All three levels of arrangement give the soil its structure and result in a series of

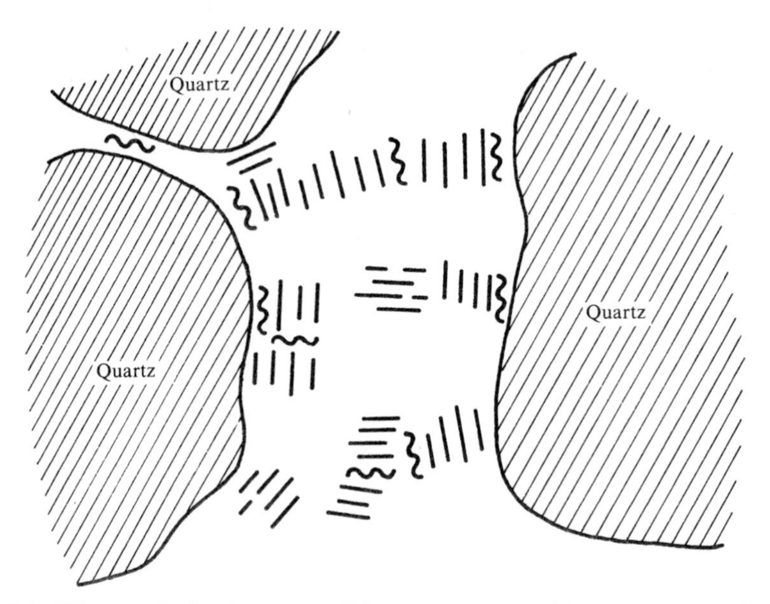

Fig. 2.1. Diagram indicating a possible arrangement of large quartz and other primary minerals, groups of clay crystals (straight lines) and organic matter (wavy lines) to form a micro-aggregate. Each group consists of clay crystals sufficiently close to act as a single unit and all combinations of orientation and relative distribution are possible. Redrawn after Emerson, W.W. (1959). *J. Soil Sci.* **10**, 235–44.

interstices, ranging from the inter-lattice spaces of the clay crystals, through the intra-aggregate spaces as visualized in Fig. 2.1 and inter-crumb spaces, to worm-holes, root channels and cracks and fissures. There is, therefore, an interconnected network of irregularly shaped pores, varying greatly in size and occupied by air and water.

This pore space is important to allow the penetration of roots – as well as the entry of water and air. It results in the soil having a much lower apparent (or bulk) density (weight per unit volume of soil *in situ*) than would be expected from the true density of the constituent minerals (about 2.65 g cm^{-3}, although some components have densities up to 5 g cm^{-3}) and organic matter (1.2–1.7 g cm^{-3}). If the soil consisted of uniform spherical particles, the tightest packing would give a porosity (volume of air and water per unit volume of soil *in situ*) of about 26 per cent. Actually, the pore space varies from about 40 to 60 per cent in most agricultural soils (Table 2.1) but may vary widely even in the same soil with changes in degree of aggregation. The apparent density varies from 1 to 2 g solids per cm^{-3} moist soil depending on its composition, degree of aggregation and moisture content. Generally, the porosity decreases and the apparent density increases with depth.

Table 2.1. Porosity and apparent density of typical soils

	Porosity (%)	Apparent density (g cm^{-3})
Sandy soil	35	1.7
Loam	46	1.4
Heavy clay	53	1.2
Clay with much organic matter	60	1.0

No really satisfactory data are available concerning the size distribution of pores in soils. However, we have put together some figures from different sources in Table 2.2 which we believe are reasonably approximate. Although water supply will be discussed later, it is convenient here to classify pore sizes in the classes distinguishing (i) those which are too large to hold water against gravity ($> 30\ \mu m$ diameter) from (ii) those which hold water too tightly for plants to extract ($< 0.2\ \mu m$ diameter) and (iii) the intermediate group which hold the water generally available to plants. Some of the clay soils may show distributions appreciably different from that illustrated depending on the degree of aggregation, flocculation and cracking.

Roots of annual plants have diameters somewhere in the range of 5×10^{-2} to 5×10^{-3} cm with root hairs of, say, 1×10^{-3} cm. Even in the upper soil layers of highest root concentrations, roots rarely occupy more than 0.3–1.5 per cent of the soil volume – with higher concentrations in sands than in clay soils. Some evidence suggests that roots cannot penetrate rigid pores less than about 200 μm in diameter nor through systems with shear strength greater than 3×10^3 kg m^{-2}; the shear strength of many soils in bulk is greater than this. It seems likely then that many parts of soil aggregates are not occupied by roots, and that most of the largest interstices are occupied by them. This contributes to their unequal distribution through the soil, particularly at a microlevel; nevertheless, there are few soils through which a root cannot find some way. It is only peculiar occurrences, such as a 'hard pan' or the parent rock material, which restrict the root growth of

Table 2.2. Pore-size distributions in three types of soil

Category		Proportion of total volume in:		
		Coarse sand	Clay loam	Heavy clay
Solids		0.68	0.44	0.38
Water- and air-filled pores of diameter	$> 30\ \mu m$	0.24	0.10	0.04
	$0.2\text{–}30\ \mu m$	0.07	0.27	0.25
	$< 0.2\ \mu m$	0.01	0.19	0.33

annuals to depths less than the 1–2 m they normally penetrate. There is also an unequal distribution in depth, the vast proportion being in the upper 20 cm.

Supply of mineral elements from the soil

The inorganic elements used by plants are present in the soil as ions in the soil solution, as ions adsorbed on the clay minerals and organic matter, in readily decomposable inorganic compounds, in very stable minerals, in the decomposing organic matter and in the living micro-flora and -fauna. Those immediately available to the plant are present as cations and anions in the soil solution; the solution in turn is in equilibrium with the absorbed ions (mainly cations) and as the soil solution loses ions to the plants these are, at least in part, replenished. It is therefore no easy matter to measure the amounts of the different nutrients that may become available to a crop during its growth. Because the vast proportion of each of the elements, including nitrogen, is contained in relatively stable compounds, a total analysis is useless. Displacement of the soil solution and its analysis indicates only what is immediately available at that point of time. What are needed are estimates of the amounts present in the soil solution at the beginning, i.e. at sowing, of the amounts potentially available during the growing season, and assessments of the various factors which influence this release. It is usually sufficient to make measurements in the top 20–30 cm layer of soil as it is here that nutrients, especially in artificially fertilized soil, are in highest concentration and where most of the roots develop.

There are fairly large differences in behaviour between the different nutrient elements and it is best to consider these individually. We will be concerned only with N, P, K, Ca and Mg, neglecting in this elementary treatment elements such as Fe, Mn, S, Zn, Cu, Mo and B which are equally essential but are required in much smaller amounts.

Nitrogen

A simplified version of the nitrogen cycle in soils is shown in Fig. 2.2. Effective additions of nitrogen to the soil are made by fertilizer applications, in rain (0.2–2 g m^{-2} y^{-1}) or by the symbiotic bacterium *Rhizobium* associated with legumes (up to 10 g m^{-2} y^{-1} by a pure clover or lucerne crop). Free-living N-fixing organisms, denitrifying organisms, and losses by volatilization may be ignored, as their contributions are too small to be significant within the time-scale of a growing season, the period with which we are concerned. Losses from the soil occur by plant uptake or by leaching beyond the root zone. Nitrogen is taken up by plants in the form of either NH_4^+ or NO_3^- ions; the former are usually adsorbed on the clay minerals and humus and the latter free in the soil solution. Hence, heavy falls of rain which replace all of the water in the root zone will wash all NO_3^- ions into the lower layers.

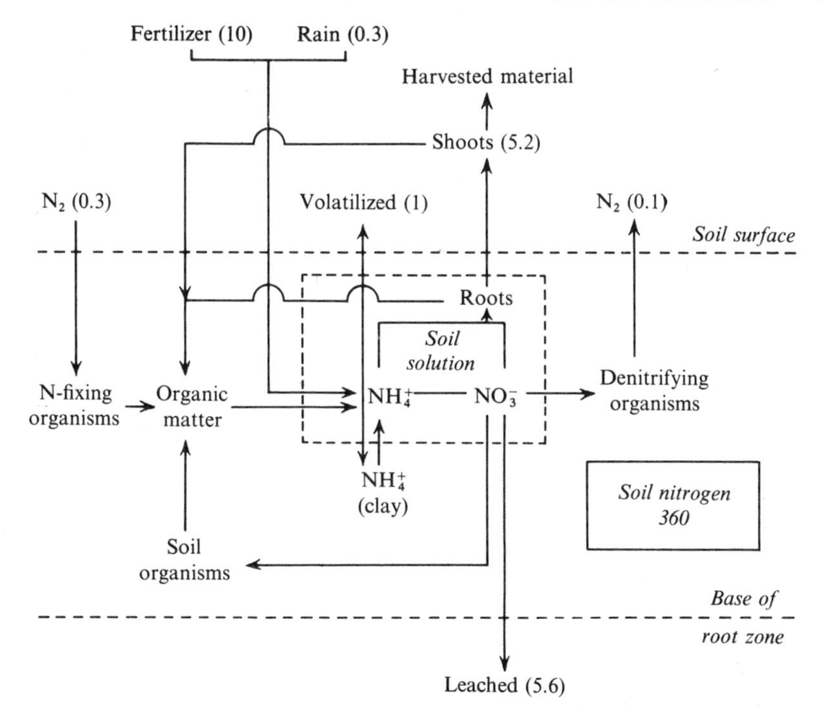

Fig. 2.2. Diagram illustrating the nitrogen cycle. The figures indicate the annual changes (g m^{-2}) in the top 22 cm of a cropped soil at Rothamsted, England. Losses by volatilization and denitrification are always small; gains by free-living N-fixing organisms are always small but those by the *Rhizobia* associated with legumes may add up to 10 g m^{-2} y^{-1}. The other components vary appreciably but the rate of turnover is always small compared with the total amount of nitrogen present.

The $NO^-{}_3$ ions are being constantly replenished via $NH_4{}^+$ from decomposing organic matter and are being removed, as are $NH_4{}^+$ ions, by absorption by plants and the many soil organisms. $NH_4{}^+$ ions are usually converted to $NO_3{}^-$ fairly rapidly by *Nitrosomonas*, *Nitrobacter* and possibly other bacteria.

The concentration of ions in the soil solution and adsorbed on the clay minerals, i.e. those readily available to plants, depends on the rates at which each of the contributing processes proceeds. These vary appreciably in different circumstances. The annual balance sheet shown for one set of circumstances in Fig. 2.2 gives a crude overall picture and cloaks important short-term changes. However, it draws attention to two significant features: that only about 1 per cent of the nitrogen present in the soil is removed by a crop and rarely is more than half to three-quarters of the nitrogen added as fertilizer recovered by the following crop.

Let us explore in a little more detail some of the changes which occur in a range of situations. Suppose we start with arable soil immediately after a crop of wheat has been harvested. We might expect then to find about 1 mg N as NH_4^+ and 2 mg N as NO_3^- per kg dry soil (i.e. 3 ppm) – say about 0.3 g m^{-2} in the top 25 cm. The total N in this layer is likely to be 200–400 g m^{-2}; i.e. a thousand-fold more. A grass–clover ley is then sown and maintained for, say, 5 years. During this time, the changes in NH_4- and NO_3-N would be small, N as NO_3^- staying about 1 ppm and N as NH_4^+ slowly increasing to about 3 ppm. During this time all of the processes shown in Fig. 2.2 are operating, but the NO_3^- and NH_4^+ ions are being absorbed by the plant and soil organisms as rapidly as they form. Estimates of total N would show that this is increasing by about 4–10 g m^{-2} y^{-1}. Ignoring amounts removed from the area as animal products or hay, which can be readily determined, we would, after 5 years, have about 340 g m^{-2}. The land is then ploughed up and left as bare fallow. There is then a fairly rapid rise in NO_3-N, say at a rate of 0.3 ppm per day. (The rate of mineralization of nitrogen, as it is called, is a function of the total organic-N content, C_N g (g soil)$^{-1}$, and the temperature, T (°C), and is approximately given (in g d^{-1}) by $10^{-7}(0.2T-1)C_N$, provided the soil water is in the 'available' range.) The environment is one with frequent light showers of rain so that the nitrifying bacteria continue to work but the NO_3^- ions are not leached; falls of rain of about twice the water-holding capacity of the soil (i.e. about twice 100, 200 and 300 mm per m depth of sands, loams and heavy clays, respectively) will leach out most of the NO_3^- ions then present. After about six months' bare fallow, there is then about 60 ppm or 18 g m^{-2} N as NO_3^- present. (Analyses of arable soils show values of 10–60 ppm, rising to 100 ppm with tropical soils in the dry season.) This together with added nitrogen fertilizer is then available for the growth of the crop. If green manure or other organic matter is added to the soil, this will lead to a flush of growth of micro-organisms and absorption of the available nitrogen by them; this gradually becomes available again as the organic matter is decomposed.

Phosphorus

Phosphorus occurs in the soil as PO_4 groups in certain minerals, as insoluble calcium phosphates, and adsorbed on the surfaces of hydrated iron and aluminium oxides. It also occurs in organic compounds, principally as inositol phosphates and nucleic acids. It seems reasonably certain that it is absorbed by plants only as the $H_2PO_4^-$ ion. Supply to the plant depends on the equilibrium and rates of release of $H_2PO_4^-$ ions from the calcium, iron and aluminium compounds to which it is bound and from decomposing organic matter. When the soluble form of monocalcium phosphate $Ca(H_2PO_4)_2$ is added, as superphosphate, to the soil, it rapidly becomes converted to the dicalcium form $CaHPO_4$, which is adsorbed on the clay minerals. This will continue to maintain a concentration of $H_2PO_4^-$ ions in the soil solution of 0.003 to 3 ppm depending on the soil type (cf. Table 4.1). However, the dicalcium form is converted to more insoluble forms which are only very slowly available. Thus, it is usual to find that rarely more than 10 per cent of the phosphorus added as fertilizer is recovered in the first crop, and perhaps another 10 per cent may be regained in a second crop.

Much of the chemistry of phosphorus in the soil is still very uncertain and there are as yet no completely satisfactory means of determining the amounts which are available; account must be taken of the rates of immobilization into insoluble forms as well as of the equilibrium constants between these and $H_2PO_4^-$. Applying a large degree of simplification, the soil phosphorus may be regarded as being in three forms, all of which are in equilibrium:

$$\text{Solution-P} \rightleftharpoons \text{labile-P} \rightleftharpoons \text{non-labile-P}$$

Solution-P possibly consists mainly of $H_2PO_4^-$ ions in the soil solution; labile-P as ions or reasonably soluble crystals such as hydroxylapatite adsorbed on clay minerals; and non-labile-P as that contained in effectively insoluble compounds. The rate at which one form changes to another depends on the equilibrium constants and the amounts of reactants; these vary appreciably between different soils but apparently cannot yet be predicted from other soil properties. However, they can be determined empirically with acceptable precision, as will be explained in Chapter 4. The reaction constants of the first group would be measured in days and that of the second in months; hence, in the current context, the phosphorus status involves only the first two forms. Nevertheless, as mentioned above, up to 80 per cent of added phosphate may be converted into non-labile forms.

Potassium, calcium and magnesium

These elements occur mainly as cations adsorbed on the surfaces of clay minerals and various compounds in the organic matter. They also occur in a number of the soil minerals; the potassium-bearing minerals, in particular, yield K^+ ions fairly readily, but at rates which vary with different species of plants. Some soils adsorb and retain K^+ fairly tenaciously; hence, recoveries of added fertilizer vary from 30 to 100 per cent depending on the soil. The adsorbed ions are exchangeable with other ions, especially H^+ ions from plant roots; the ease of replacement of ions on the clay minerals being usually in the order Na > K > Mg > Ca > H. The supplying-power of the soil, and hence the fertilizer requirement, can often be gauged from the content of exchangeable ions, although this is a poor criterion in soils undergoing rapid weathering, and values (meq kg^{-1}) from a variety of soils lie within the ranges 11–300 for Ca^{++}, 7–50 for Mg^{++}, 8–17 for K^+ and 5–70 for Na^+. The overall content of exchangeable cations in a soil and the ratios of the individual ions depend on the conditions under which the soil has developed, treatments, such as liming, it has received, and the relative amounts of the different clay minerals and humus it contains – the exchange capacity (meq kg^{-1}) of humus being 1500–3000, montmorillonite 1000, illite 300 and kaolinite 100.

Soil water

A given volume of soil V consists of a volume of solids V_s, a volume of water V_w, and a volume of air V_a, of mass M_s, M_w and M_a, respectively. We have already defined (p. 8) the *porosity* as $(V-V_s)/V$ and the *apparent density* as M_s/V. The *water content* θ may be defined on a volume basis as V_w/V or, more usually, on a weight basis as M_w/M_s, where M_s is taken as the mass after drying to equilibrium at 105 °C. The former may be obtained from the latter as $\rho_s M_w/(\rho_w M_s)$ and expressed as the equivalent depth of water in unit depth of soil, i.e. the depth the water in the soil would have if impounded on a horizontal impervious surface. This is a very convenient expression when considering evaporation.

The water content of a soil is not a reliable index of the availability of the water to plants. A more useful concept is that of the soil water potential, or amount of work required to move a given amount of water from one point to another compared to that required to move the same amount of pure free water the same distance. In formal terms, the water potential ψ is defined as 'the amount of work that must be done per unit quantity of pure water in order to transport reversibly and isothermally an infinitesimal quantity of water from a pool of pure water at a specified elevation at one standard atmosphere pressure to the soil water (at the point under consideration)'. This is a very convenient term as it reflects the capability of the water to do work compared with free pure water and thus removes difficulties of measuring absolute energy states. Pure water is taken as zero and, because the free energy of a liquid is lowered by capillary forces or dissolved substances, the water potential of soil water is always negative. Water will always tend to move from a higher to a lower potential. The water potential is usually expressed in energy units per unit mass, volume (i.e. units of pressure) or mole. (Note 10^6 erg g^{-1} = 10^2 J kg^{-1} ≡ 0.987 atmosphere = 1 bar ≡ 1.017×10^3 cm water).

Three components contribute to the soil water potential. One, the *osmotic potential* π, arises from the presence of solute particles. In dilute solutions it is effectively independent of the kind of solute particle, be it an ion, a small undissociated non-electrolyte or a large particle; only the number is important. The relationship is conveniently expressed by

$$\pi \simeq -RTC_s \qquad 2.1$$

where C_s is the molal concentration ($C_s = 1$ with 1 mole undissociated solute per kg water), R the gas constant (= 8.315 J °K^{-1} $mole^{-1}$) and T the temperature (°K). The solute particles influence the energy status of the water but not its movement in response to applied pressures or gravitational forces; they can, however, affect the movement of the water across differentially permeable membranes.

The second component influencing the soil water potential is the *matric potential* τ, which is related to the absorptive forces of the soil matrix. The most strongly held water is that held to the clay and humus colloids by hydrogen bonding. As additional water layers accumulate around these, the force with which each successive layer is held diminishes, but it is not until several such layers are present that the outermost is held by a force sufficiently small to be absorbed by plants. Water in this strongly adsorbed state is of little concern to us; however, it does explain the differences between the amounts of non-available water held by different soils (cf. Fig. 2.5).

Further amounts of water added to the soil are held more loosely, occupying the various interstices in the matrix. Surface tension forces now predominate, the force with which the water is held being inversely proportional to the diameter of the pore. (There are, however, secondary influences associated with the swelling and shrinkage of the soil on wetting and drying.) Taking a completely saturated soil, then, differing proportions of the water can be removed with differing degrees of ease (i.e. amounts of work), this distribution being closely related to the distribution of pore-sizes in the soil. The relation between the change in free energy of the water and the diameter (D μm) of the pore in which the water is held is given by

$$\tau = -4\sigma/\mathrm{D}\ \mathrm{dyne\ cm^{-2}} \simeq -2.9\times10^{2}/\mathrm{D\ J\ kg^{-1}} \qquad 2.2$$

where σ is the surface tension of water.

Thirdly, a pressure potential P may be recognized, taking into account the pressure due to a head of free water *above* that being considered. That is, P of water below a water table in saturated soils is positive and becomes zero in unsaturated soils, τ is zero in saturated soils and becomes negative in unsaturated soils, and π is always negative. These three components are additive, and hence the water potential is given by

$$\psi = \pi+\tau+\mathrm{P} \qquad 2.3$$

The water characteristics of any soil are best defined by the curve relating the matric potential to the moisture content (Fig. 2.3 and cf. Fig. 2.4). Only a portion of the water is readily available to plants. That held when all pores with an effective diameter not exceeding about 30 μm are filled is known as the field capacity (τ about -10 J kg^{-1}); this is a rather indefinite point representing the water remaining after free drainage has ceased. That held when the bulk matric potential is about -1500 J kg^{-1} is called the wilting point; this approximates to the water potential developed by most crop plants at wilting but, again, it is a very variable value. The water contained between these two values provides a crude estimate of that normally

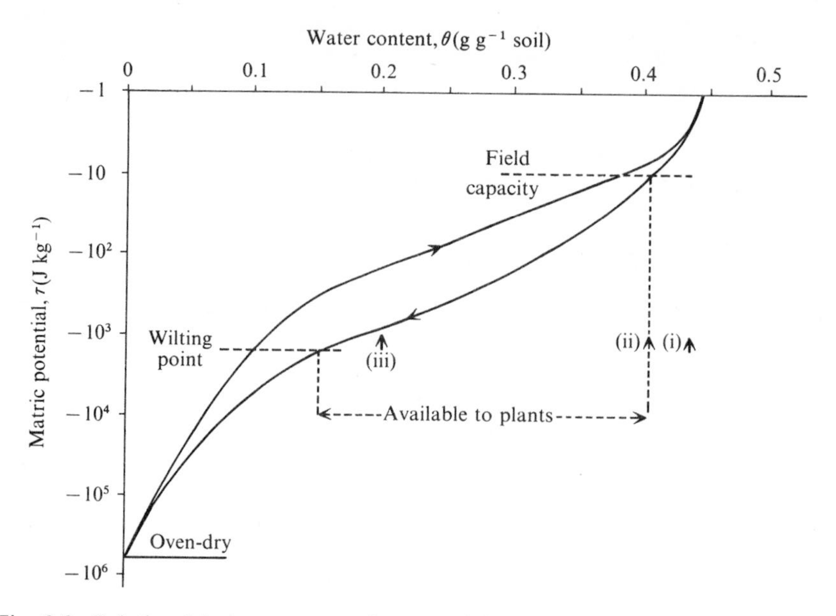

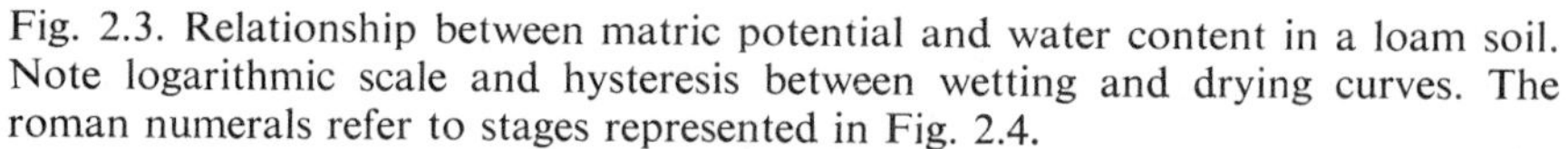
Fig. 2.3. Relationship between matric potential and water content in a loam soil. Note logarithmic scale and hysteresis between wetting and drying curves. The roman numerals refer to stages represented in Fig. 2.4.

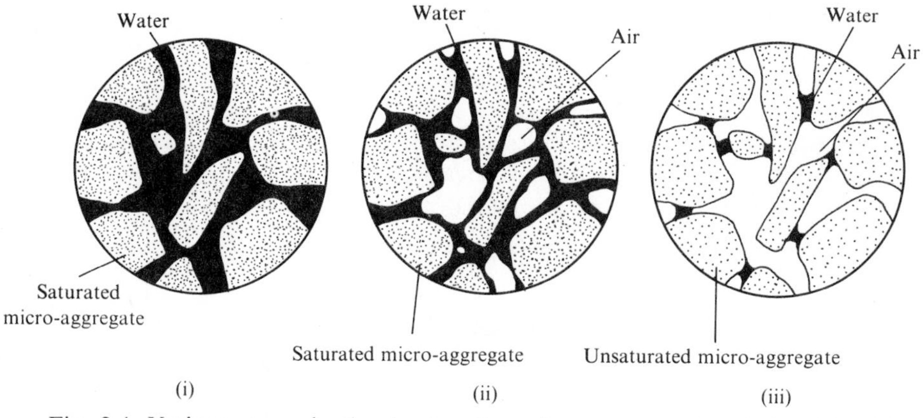

Fig. 2.4. Various stages in the desaturation of an aggregated soil. See Fig. 2.3. After Youngs, E. G. (1965), *Symp. Soc. exp. Biol.* **19**, 89–112.

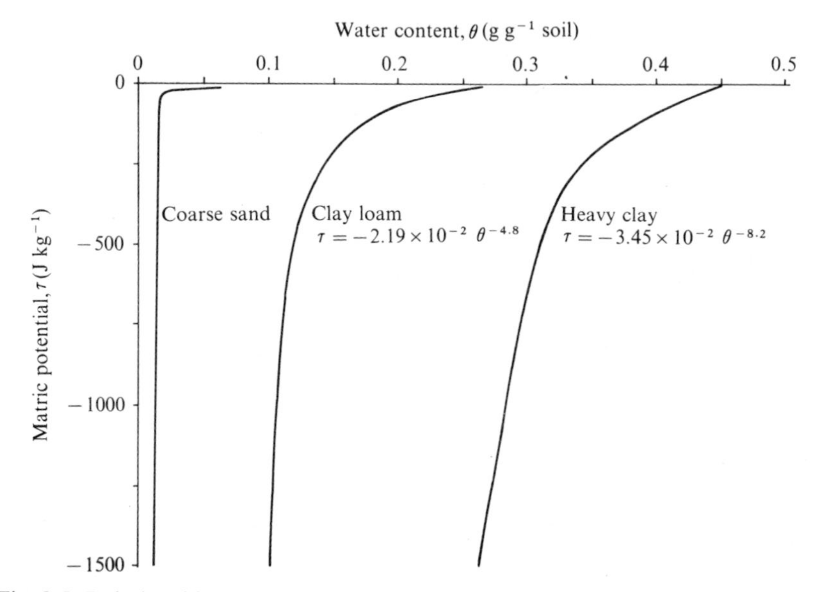

Fig. 2.5. Relationship between matric potential and water content over the available range in three different types of soil.

available to plants. It will be noted that the curves are somewhat displaced in position according to whether the soil is being wetted or dried, i.e. there is a hysteresis effect. The moisture characteristics as defined by these curves vary appreciably between soils. Of the three examples given (Fig. 2.5), the amount of available water held in the coarse sand, loam and clay soils, respectively – when fully replenished – was 0.06, 0.19 and 0.40 g (g soil)$^{-1}$; assuming apparent densities of 2.0, 1.4 and 1.2 g cm^{-3}, this gives storage capacities of 120, 196 and 233 mm per metre depth of soil. The curve can often be described sufficiently by $\tau = -a\theta^{-b}$, where a defines the position and b the degree of slope of the curve. Values are shown on Fig. 2.5 for the curves relating to the loam and clay soils but that for the coarse sand could not be so described.

Soil atmosphere

The atmosphere in the soil usually contains about ten times the concentration of carbon dioxide, twice the concentration of water vapour and less oxygen and nitrogen than in the atmosphere proper (Table 2.3 and p. 20). These concentrations vary of course with the activity of micro-organisms, soil fauna and roots, the respiration of which leads to an inward flux of oxygen and an outward flux of carbon dioxide. This activity, together with the impedance to diffusion of gases within the soil, prevents the maintenance of any equilibrium. Most of the exchange occurs by molecular diffusion, and

Table 2.3. Approximate composition of atmosphere proper and soil atmosphere (per cent by volume)

	N_2	O_2	CO_2	Water vapour
Atmosphere proper	78.2	20.7	0.03	1.0
Soil atmosphere	77.6	20.2	0.2	2.0

is adequately described by Fick's law where the coefficient of diffusion in soil is given by $0.66SD$, D being the diffusion coefficient of the gas in free air, S the fractional soil volume filled with air, and 0.66 a measure of the increased length of the diffusion path owing to the tortuosity of the pores. That is, the diffusion coefficient is effectively zero in soil completely saturated with water and increases as the soil dries towards the wilting point to about one-fifth of that appropriate to free air.

The measured rate of carbon dioxide evolution varies between about 2 and 20 $g\ m^{-2}\ d^{-1}$, depending on temperature, presence of vegetation and water content. As a first approximation, it may be taken as $3^{T/10}$ and $3^{T/10}+5\ g\ m^{-2}\ d^{-1}$ ($5\ °C < T < 25\ °C$) from moist bare and cropped soil, respectively. Despite these rates of production and relatively slow rates of exchange, it is unlikely that the roots of crop plants are ever seriously depleted of oxygen except during prolonged periods of water-logging. The effects which arise then appear to be due more to secondary processes producing toxic substances than directly to anoxia.

Soil temperature

The temperature fluctuations in the soil are described empirically by two periodic fluctuations – one diurnal (Fig. 2.6) and one seasonal (Fig. 2.7) – which vary with depth. These approximate to sine waves, the largest fluctuation (greatest amplitude) occurring at the soil surface. As the wave passes down into the soil, its amplitude is damped enormously and the time lag of the oscillation, referred to that at the surface, increases. Diurnal fluctuations decrease exponentially with depth and at 35 cm are about 5 per cent of those at the surface. They are also delayed, the maximum being achieved at about the same time as the surface minimum. Similarly, the maximum seasonal temperature at 3 m is reached in late autumn whereas that at the surface is attained in mid-summer; the amplitude of the annual wave at a depth of about 6–7 m is about 5 per cent of its surface value.

These changes arise from the rates of supply of heat to and from the soil surface and its transfer within the soil. The soil surface receives heat as short-wave radiation of which 10 (wet soil) to 20 (dry soil) per cent is immediately reflected (cf. p. 22). The remainder is absorbed; part of it is

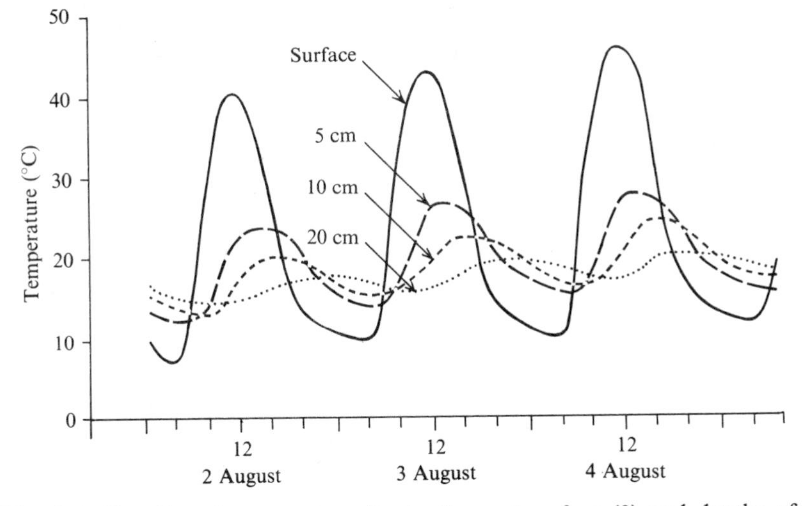

Fig. 2.6. The diurnal temperature fluctuations at surface (0) and depths of 5, 10 and 20 cm in bare soil on three successive clear summer days. After Russell, E. W. (1953). [*Note*: where an acknowledgement is cited in this abbreviated form the full reference will be found in the 'Further Reading' list.]

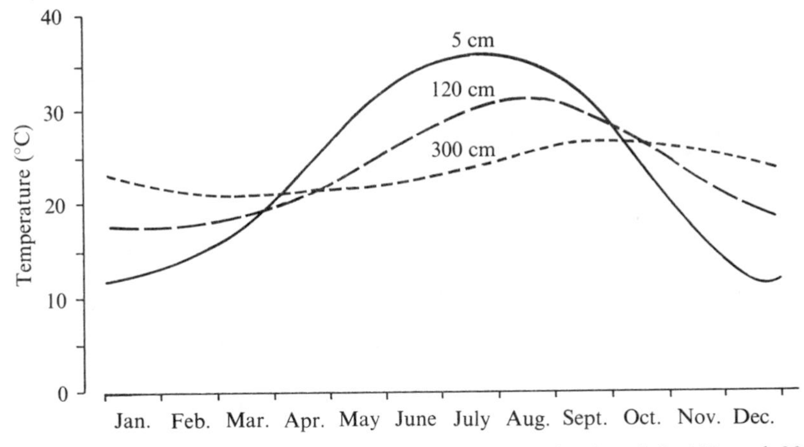

Fig. 2.7. Annual temperature waves in bare soil at depths of 5, 120 and 300 cm (Northern Hemisphere).

re-radiated into the atmosphere, part conducted to the atmosphere above, part used to evaporate water, and part conducted into the soil. The surface temperature is dominated by the fluxes of incoming radiation and re-radiation and, hence, shows the greatest variation and is much influenced by the degree of cloud. If the surface layers are wet, much of the energy absorbed is used in evaporation. Dark surfaces absorb and emit radiation more efficiently than do light-coloured surfaces, and so dark soils show wider fluctuations than light-coloured soils. The surface temperature, and that of the lower layers, depend also on the rate of heat conduction through the soil and its heat capacity. These properties vary mainly with the water content. On vegetated surfaces, the fluctuations in temperature of the surface and deeper layers are much less than with bare soils, as much of the heat exchange occurs in the leaf canopy rather than at the soil surface.

As a broad generalization, about 10–30 per cent of the instantaneous net radiation may be absorbed during the day while during the night the soil contributes a higher proportion of the out-going radiation. There is a net gain of heat by the soil equivalent to about 2 per cent of the incoming radiation during the summer and much the same net loss during the winter.

THE AERIAL ENVIRONMENT

The atmosphere is a fluid composed of (i) clean dry air (comprising 78.08 per cent nitrogen, 20.95 per cent oxygen, 0.93 per cent argon, 0.03 per cent carbon dioxide and traces of other gases, and with a mean molecular weight of 28.9); (ii) water vapour (up to 0.6, 1.2, 2.2 and 3.8 per cent by volume at saturation at 0, 10, 20 and 30° C, respectively; the saturated vapour pressure e' is related to temperature over the range 0–50 °C by

$$e' = 6.108 \exp\{17.2674T/(T+237.28)\};$$

and (iii) various impurities such as particles of smoke, dust and pollen, and industrial and other man-made gases and their transformed products (including peroxyacyl nitrates and similar oxidants).

The climate in the few metres above the ground is determined by the large-scale circulation, arising initially from unequal heating of the earth's surface and modified by Coriolis effects, and by the local flux of solar radiation. For our purposes, we can simply accept the former but need to consider some features of the radiation balance.

Radiation

The average radiant flux density reaching the outside surface of the atmosphere facing the sun is about 1390 W m^{-2}. Mean daily fluxes at the different latitudes throughout the year are given in the *Smithsonian Meteorological Tables*. On passing through the atmosphere, some is absorbed, particularly

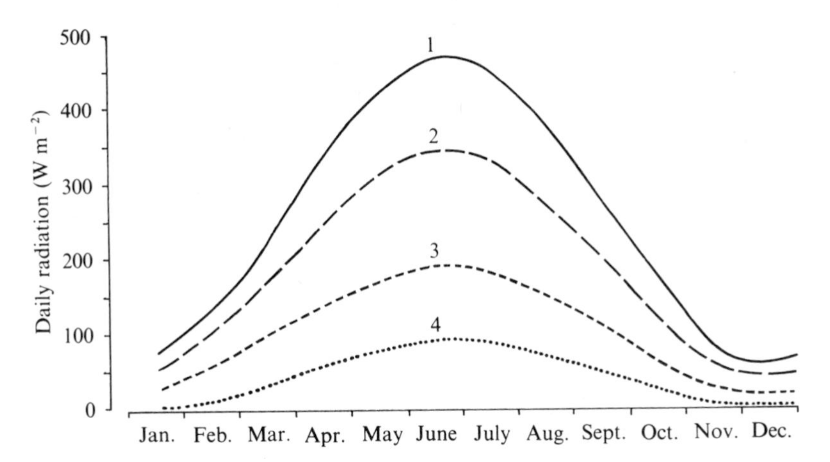

Fig. 2.8. Variation in daily radiation at latitude 55 °N. (1) Radiation at outside of atmosphere; (2) expected radiation if all days were cloudless; (3) recorded radiation (10-year average); (4) expected radiation if all days were cloudy.

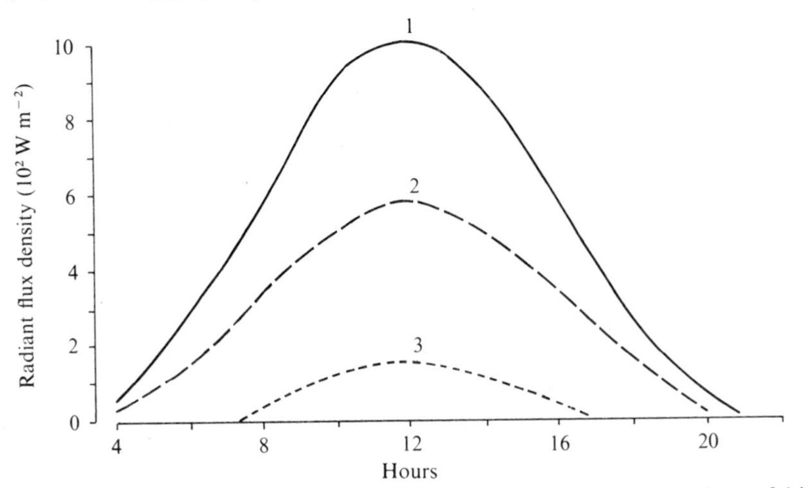

Fig. 2.9. Variation in diurnal radiation at latitude 55 °N. (1) On days of highest radiation; (2) average during summer; (3) average during winter.

by clouds, and some is scattered by dust and other constituents. The portion reaching the surface, the short-wave *global radiation*, Q_S, therefore has two components – one direct and one diffuse. With clear skies the diffuse component comprises some 10–15 per cent; on cloudy days all is diffuse. On a completely clear day about 75 per cent, and on a completely overcast day about 25 per cent, of the radiation reaching the outside of the atmosphere penetrates to the earth's surface. About half of this global radiation is in the 0.4–0.7 μm wave-length band (i.e. photosynthetically active) and about

Table 2.4. A daily radiation balance on a clear summer day

	$W\,m^{-2}$
Income	
Incoming energy outside atmosphere	363
Incoming energy at surface	254
Reflected	51
Outgoing radiation	85
Net radiation at surface	118
Use at surface	
Evaporation	101
Heating air and plants	10
Heating soil	5
Plant growth	2
	118

half in the 0.7–3 μm band. Seasonal and diurnal variations are illustrated in Figs. 2.8 and 2.9. Over periods of a week or more, an average value of Q_S can be estimated from $Q_S = Q_A\ (0.25+0.51n/N)$ where Q_A is the amount reaching the outside of the atmosphere and n and N are the hours of bright sunshine and daylight, respectively.

About 7 per cent of the photosynthetically active radiation (light) and 30–40 per cent of the infra-red radiation is reflected from leaves; i.e. about 25 per cent of the *total* radiation is reflected from vegetation compared with 10–20 per cent from bare soil and about 5 per cent from water.

There is also continuous exchange of long-wave radiation (about 5–30 μm wave-length) between the earth and the atmosphere, the flux from the sky being given by $\beta\sigma T^4$ and that from the earth by σT^4, where σ is the Stefan–Boltzmann constant and β an emissivity which varies from about 0.7 with clear dry skies to near unity in humid cloudy weather. With the same temperature, T(°K), of sky and earth, the net upward flux,

$$Q_L = (1-\beta)\sigma T^4 \qquad 2.4$$

where $1-\beta \simeq (0.56-0.08e^{0.5})(1-0.9c)$, the vapour pressure e being in mbar and c being the fraction of the sky covered by cloud. That is, with clear skies, the net outgoing radiation is ten to thirty times that found under completely overcast conditions.

Equating the above components,

$$Q_N = Q_S(1-\alpha)-Q_L \qquad 2.5$$

where α is the reflection coefficient and Q_N the *net radiation*, i.e. that absorbed at the surface and transformed into other forms of energy. These are latent heat (evaporation of water), sensible heat (warming up the lower layers of the atmosphere, the vegetation, and the surface layers of the soil), and chemical energy (fixed in products of photosynthesis). Although the proportions vary appreciably in different situations, Table 2.4, relating to a vegetated surface on a clear summer day and with ample water available, gives a crude approximation of the fate of the energy.

Interception by vegetation

A crop may be regarded as consisting of a number of horizontal layers of leaves (more strictly, laminae), each of unit leaf area per unit area of soil, L. The distribution of these layers with height above the soil surface is far from regular, the narrowest layer – i.e. the region of greatest density – often being near the centre or even towards the top of the crop (Fig. 2.10). The maximum density of laminae (area per unit volume), L_V, may vary from about 1 in a thick crop of clover, through 0.02 $cm^2 cm^{-3}$ in a crop such as maize, to zero in lower layers where the leaves have senesced; rarely is more than 1 per cent of the total canopy volume occupied by plant tissue. Where the plants are reasonably uniformly distributed, for example, a cereal crop near ear emergence, the leaves are more or less randomly distributed in the horizontal plane. The angle which each lamina or portion of each lamina makes with the horizontal varies appreciably with the species and the stage of growth; in crops such as clover, kale and beans there is a preponderance of small angles, whereas the younger leaves of grasses are more erect and become 'flatter' as they age. Therefore, at a young stage or towards the top of a stand of grass, there is a large proportion of large angles; the distribution becomes more uniform as the stand ages or in the lower layers of a thick canopy.

Two components of the flux density of direct and diffuse radiation which reaches the outside of the canopy are relevant, and it is convenient to consider both together. One is the incoming flux of visible radiation which is significant in photosynthesis. The other is the flux of total incoming short-wave radiation which, together with the net out-going flux of long-wave radiation, provides the net radiant energy available for evaporation. A complete analysis of the fate of the radiation within a crop stand is well beyond our terms of reference; such a treatment has been provided by Cowan (cf. Fig. 3.5) and a less thorough and laborious analysis is given by Monteith (1973, and in Eastin *et al.* 1969). We will generally follow Monteith, but in a prescriptive rather than an explanatory way.

Consider the flux, Q_0, of radiation of a specified spectral composition received at the top of the crop. Some of this passes down through gaps between the leaves and some falls on leaves; of the latter a proportion, α, is reflected and a proportion, γ, transmitted. The attenuation of the radiation as it passes down through the canopy can be measured by long horizontal tube-solarimeters giving an average value over a plane, above which there is leaf area, L. Usually, the flux on any plane can be represented by

$$Q = Q_0 \exp(-kL) \qquad 2.6$$

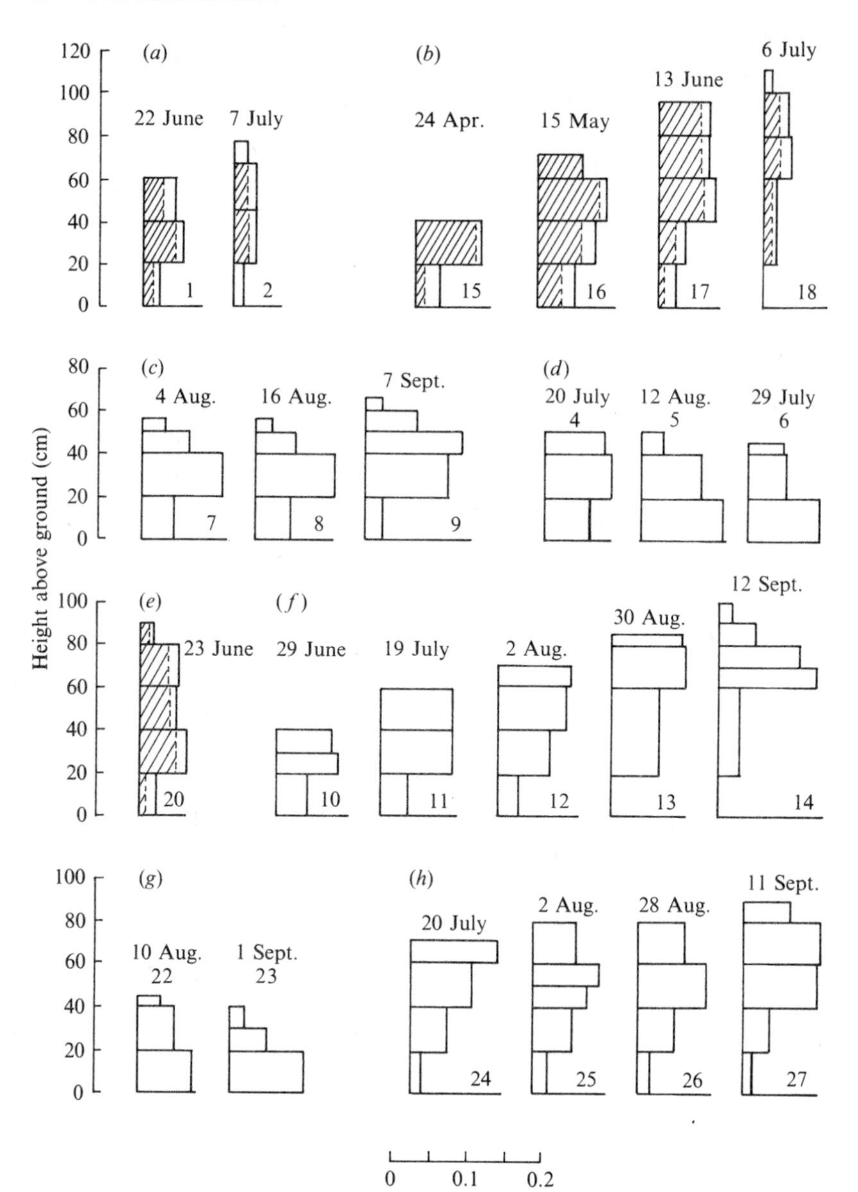

Fig. 2.10. The vertical distribution of leaf area density, L_v, in a number of crops. The leaf area within each layer (cm^2 leaf per cm^3 canopy volume) is given by the horizontal dimension and the depth of the layer by the vertical dimension. (*a*) 1, 2, barley, 1960; (*b*) 15–18, winter wheat, 1961; (*c*) 7–9, sugar beet, 1960; (*d*) 4–6, potato, 1960; (*e*) 20, spring wheat, 1961; (*f*) 10–14, kale, 1960; (*g*) 22–23, potato, 1961; (*h*) 24–27, kale, 1961. Crops grown at Rothamsted, England. After Leach, G. J. & Watson, D. J. (1968). *J. appl. Ecol.* **5**, 381–408.

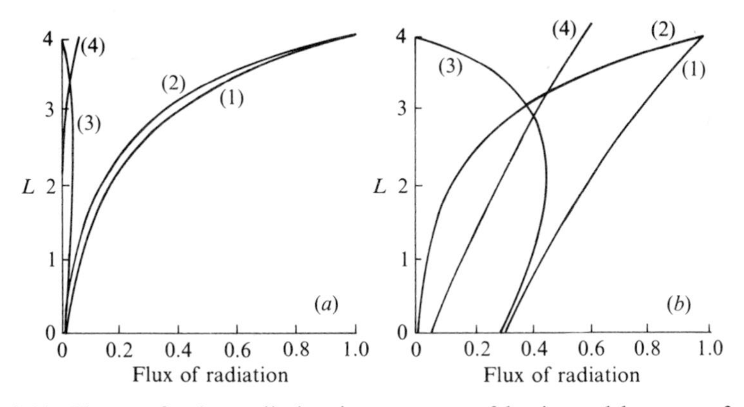

Fig. 2.11. Fluxes of solar radiation in a canopy of horizontal leaves as functions of cumulative leaf area, L, relative to the downward flux of radiation at the top of the canopy. (1) is the total downward flux of radiation; (2) is the part of this flux due to unintercepted radiation and (3) the part due to radiation scattered downwards by the leaves; (4) is the upward flux of radiation due to scattering by the leaves and reflection by the ground surface. In (*a*) the scattering coefficient of the leaves and the reflection coefficient of the ground surface are values appropriate to radiation in the waveband 0.31–0.73 μm; in (*b*) the values are appropriate to the waveband 0.73–1.2 μm. After Cowan, I. R. (1968). *J. appl. Ecol.* **5**, 367–79.

The exponent, k, represents two parameters: the transmission coefficient, γ, depending on spectral composition only, and a geometrical component, K, which varies with the distribution of leaf angles within the crop and the solar elevation. These are related by $\mathrm{K} = k/(1-\gamma)$. For most crops, $\gamma = \alpha = 0.07$ for visible radiation (0.3–0.7 μm), 0.41 for infra-red radiation (0.7–3 μm), and 0.25 for total short-wave radiation. Thus, if k is obtained by measuring total incoming radiation, $\mathrm{K} = 1.33k$; if from visible radiation, $\mathrm{K} = 1.07k$. There is also an upward flux arising from the radiation reflected by the leaves (Fig. 2.11) and it must be added to the downward flux to give the total below leaf area L. To an acceptable approximation, this is given by

$$Q = Q_0 \exp\{-\mathrm{K}(1-\gamma-\alpha)L\} \tag{2.7}$$

The amount absorbed by unit area of leaf in each layer can then be obtained directly.

Equation 2.7 refers to the short-wave (0.3–3 μm) radiation. To obtain the total energy available in each layer, the exchange of long-wave radiation must also be considered. The upward flux of long-wave radiation outside the canopy is always greater than the downward flux from the sky – by about 1.2 times with transpiring vegetation on a warm clear day. The net long-wave flux within the canopy, *measured from the top downward*, can be described approximately by

$$Q_\mathrm{L} = -(1-\beta)\sigma \mathrm{T}^4 \exp(-\mathrm{K_d} L) \tag{2.8}$$

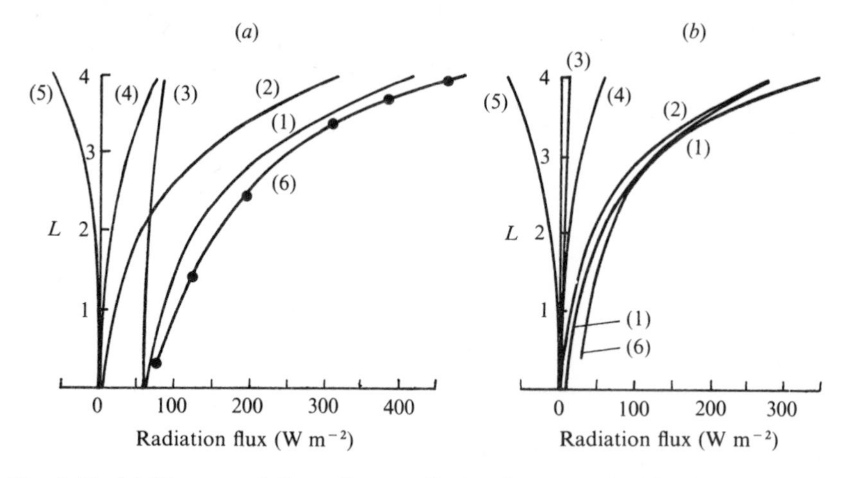

Fig. 2.12. (*a*) Downward flux of net radiation in a stand of horizontal leaves as a function of cumulative leaf area. Curve (1) is for radiation of all wavelengths, (2), (3) and (4) are the components due to solar radiation in the wavebands 0.31–0.73 μm, 0.73–1.2 μm and 1.2–5.5 μm respectively, and (5) is the component due to terrestrial radiation. Curve 6 represents mean values during a day for maize (from data of K. W. Brown).

(*b*) The net radiation absorbed by unit area of leaf in unit time as a function of cumulative leaf area, L. The conditions are the same as those for Fig. 2.12*a* and the component curves are distinguished in the same way. Both after Cowan, I. R. (1968). *J. appl. Ecol.* **5**, 367–79.

where β, σ and T are described in Equation 2.4 and K_d is the transmission coefficient, numerically equal to that for diffuse radiation through the canopy ($K_d \approx 0.7$ for foliage randomly distributed in all three dimensions, say when $K \approx 0.6$–0.7, and approaches 1 when all leaves are perfectly flat ($K \approx 1$)).

Fluxes of net radiation (Q_N) calculated for one stand of horizontal leaves (Fig. 2.12) show the behaviour with radiation of different spectral composition. Under the conditions of clear skies and a high flux of short-wave radiation, the absorption profiles of visible radiation and total net radiation are very similar. This arises mainly because of the negligible absorption in the 0.7–1.2 μm band and the fact that absorption in the 1.2–5.5 μm band is largely compensated by the net loss of long-wave radiation.

Whereas transpiration depends on the net radiation of all wave-lengths which are absorbed by each layer, the rate of photosynthesis of an area of a leaf is a function of its irradiance by visible radiation. It is necessary, therefore, to estimate the relative proportions of leaves in each layer receiving known levels of irradiance. To a reasonable approximation, this can be obtained by assuming leaves are either 'sunlit' or 'shaded' where the fractional area of sunlit leaves beneath a layer is $\exp(-KL)$ and the total area of all sunlit leaves in the canopy is its integral, i.e. $\{1-\exp(-KL)\}/K$. The flux of visible radiation, Q_V, received outside the crop, is about half of the flux of total incoming short-wave radiation, as measured by a glass-covered thermopile. It consists of direct radiation, I_0, and the diffuse component, I_d, where $I_d \approx 0.2Q_V$ on a sunny day and $I_d = Q_V$ on a completely overcast day. The total irradiance of sunlit leaves is given by the sum of:

(*a*) direct radiation from the sun, equal to $KI_0(0)$;

(*b*) diffuse radiation generated within the canopy through transmission and reflection of I_0; this is given by $K_dI_0(0)\exp(-KL)[\exp\{K(\gamma+\alpha)L\}-1]$; and

(*c*) diffuse radiation given by $K_dI_d(0)\exp\{-K_d(1-\gamma-\alpha)L\}$. The irradiance of shaded leaves is given by the sum of (*b*) and (*c*) and may be assumed to be the same for all leaves in any one layer.

Transfer processes within the lower layers of the atmosphere

The general circulation may be identified with the movement of the large-scale air masses, i.e. the weather systems represented on the daily weather charts. A variety of temporary circulations occur within these, due to unevenness in surface heating, friction, changes in local topography, and protuberances on the surface; these occur as eddies ranging in size from the planetary to those of molecular dimensions. If a very large wind-vane with a large inertia is placed at a given height, it usually stays fairly steady, indicating a general flow from one direction. If a much more sensitive wind-vane is placed nearby and the changes are followed over periods of seconds or minutes, very frequent changes of direction and velocity are seen. Similarly, such changes can be detected in the vertical dimension. If a sensitive barometer is placed at one point, very rapid fluctuations in pressure are detected. We can regard this turbulent flow as one in which small packets of air are being continually moved both upwards and downwards from one horizontal plane to another over an average distance (the *mixing length*). This leads to rapid mixing in the air layers and a much faster transfer of an entity, such as heat or water vapour, from a region of high to one of low concentration than would occur with ordinary molecular diffusion.

As the air moves over a solid surface such as the ground (or a leaf), it is slowed down by frictional forces, the layer exactly at the surface having no movement. Proceeding vertically from the surface, the successive layers have increasing velocity but these show *laminar* flow; that is, all of the air moves in the same plane as that of the surface, there being no vertical transfer between layers. Further away there is a transition region and finally the region of fully developed turbulent flow is reached. Transfer of entities, such as water vapour, within such a *boundary layer* is by molecular diffusion and is several orders of magnitude less than by turbulent transfer. Even within the turbulent region, the frictional drag of the ground is transmitted through the air, producing a shearing stress as one air layer slides over the next; this results in an associated downward flux of momentum.

Over a large expanse of a reasonably uniform surface, the transfer of any entity to or from that surface is essentially vertical; the rate of transfer of an entity across unit area of a horizontal plane (i.e. the flux density) is then proportional to the concentration gradient, the coefficient of proportionality being called an eddy transfer coefficient. In general, $F = -K\,dx/dz$, and, in particular, for the entities with which we are concerned:
Momentum (shearing stress per unit area):

$$M(\mathrm{kg\ m^{-1}\ s^{-2}}) = \rho_a K_M\, d\bar{u}/dz \qquad 2.9$$

Heat: $$H(\mathrm{J\ m^{-2}\ s^{-1}}) = -\rho_a c_p K_H\, d\bar{T}/dz \qquad 2.10$$

Water vapour: $$E(\mathrm{kg\ m^{-2}\ s^{-1}}) = -\rho_a K_W\, d\bar{q}/dz \qquad 2.11$$

Carbon dioxide: $$P(\mathrm{kg\ m^{-2}\ s^{-1}}) = -\rho_a K_P\, d\bar{C}/dz \qquad 2.12$$

where ρ_a is the density and c_p the thermal capacity of moist air, $\bar{u}$ the mean horizontal velocity, $\bar{T}$ the mean temperature, $\bar{q}$ the mean specific humidity (g water vapour per g moist air), $\bar{C}$ the mean specific concentration of carbon dioxide, K_M, K_H, K_W and K_P are the respective transfer coefficients and z the height above the surface.

The magnitudes of all four transfer coefficients are determined primarily by the degree of turbulence of the air. The velocity fluctuations in the vertical dimension increase with height above the ground; they also increase with wind speed at any given height (i.e. with the wind speed associated with large-scale air movement). K_M depends essentially on wind speed and height alone and is given by $0.4u_*z$, where u_* is a reference velocity which characterizes the particular turbulent regime, varying with the general wind speed and the surface characteristics. We cannot forecast the general wind speed associated with movement of air masses but we can measure it and also characterize the wind profile. Under neutral conditions (i.e. when frictional effects only are important, as near sunrise and sunset), this is described by $\bar{u}/u_* = 2.5\ln(z/z_0)$, where $\bar{u}$ is the mean wind speed at height z, and z_0 is an integration constant which reflects the roughness of the surface. u_* and z_0 can then be characterized by reference to the wind speed at a standard height; K_M is proportional to this wind speed and to z. For instance, for a lawn of height 1 cm and a dense wheat crop of height 50 cm respectively, z_0 was found to be 0.1 and 9 cm and u_* 26 and 63 cm s^{-1} with a wind speed of 5 m s^{-1} at 2 m height. K_M at this height would then be 2070 and 5040 cm^2 s^{-1} with the lawn and wheat crop respectively. Speeds of one-tenth and five times this would represent very calm and strong gale conditions, respectively.

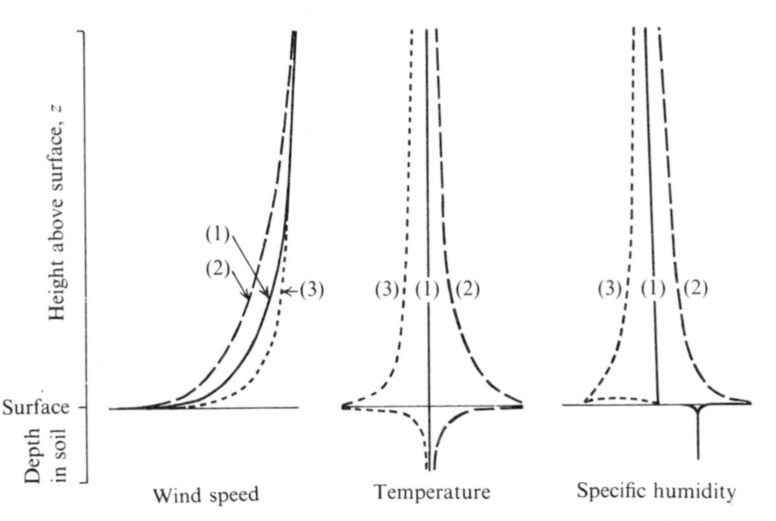

Fig. 2.13. Simplified diagram showing general nature of profiles of wind, temperature and humidity above a short smooth lawn at (1) sunrise, (2) during the middle of the day and (3) at night. The wind profiles have been adjusted to the same speed at a reference height to emphasize the changes in shape.

Even at 10 cm above the lawn under calm conditions, K_M would be 6.4 cm^2 s^{-1} or about 25 times the coefficient for molecular diffusion of water vapour in air.

Typical profiles obtained when conditions are stable (e.g. at sunset and sunrise) are illustrated in Fig. 2.13. As incoming radiation increases during the day it is absorbed most strongly by the soil or vegetation and a temperature gradient away from the ground is established. The hotter air near the surface is less dense than that in the layers above, and hence rises. This convection promotes atmospheric instability and hence increases turbulence and the transfer coefficient, so producing a more uniform temperature profile. Radiation is also used in evaporation and the humidity gradient indicates a flux of water vapour away from the surface.

At night the reverse occurs, with the soil and vegetation losing more heat than the lower layers of air, and hence there is a downwards heat flux. Under calm conditions, the heat transfer may be so slow that the cooling of the surfaces reduces the temperatures below the dew point and water condenses. Much of this dew arises from soil water.

These effects are modified by cloud (via radiation) and by wind (by increasing the forced convection). In addition, as will be discussed later, the situation becomes more complex as the height of the crop increases.

FURTHER READING

Eastin, J. D., Haskins, F. A., Sullivan, C. Y. & Van Bavel, C. H. M. (ed.) (1969). *Physiological aspects of crop yield.* American Society of Agronomy, Crop Science Society of America, Madison.

Evans, L. T. (ed.) (1963). *Environmental control of plant growth.* Academic Press, New York and London.

Monteith, J. L. (1973). *Principles of environmental physics.* Arnold, London.

Munn, R. E. (1966). *Descriptive micrometeorology.* Academic Press, New York and London.

Rose, C. W. (1966). *Agricultural physics.* Pergamon Press, Oxford.

Russell, E. W. (1953). *Soil conditions and plant growth.* Longmans, London.

Shaw, B. T. (ed.) (1952). *Soil physical conditions and plant growth.* Academic Press, New York and London.

Sutton, O. G. (1953). *Micrometeorology.* McGraw-Hill, New York.

3

The Supply and Use of Water

About 70–90 per cent of the mass of a crop consists of water. (Water contents vary with the type and age of the organ, that of air-dry seeds being about 5–15 per cent, woody stems and similar tissues 50 per cent, herbaceous roots, stems and leaves 70–95 per cent and succulent fruits 90–95 per cent.) The roots permeate a relatively wet soil, whilst the stems and leaves project into a dry atmosphere. In effect, there is a continuous flow of water from soil to atmosphere along a gradient of decreasing water potential (crudely specified in Table 3.1). On a daily basis, this flow amounts to some 1 to 10 times the amount of water held in the plant tissues, 10 to 100 times the

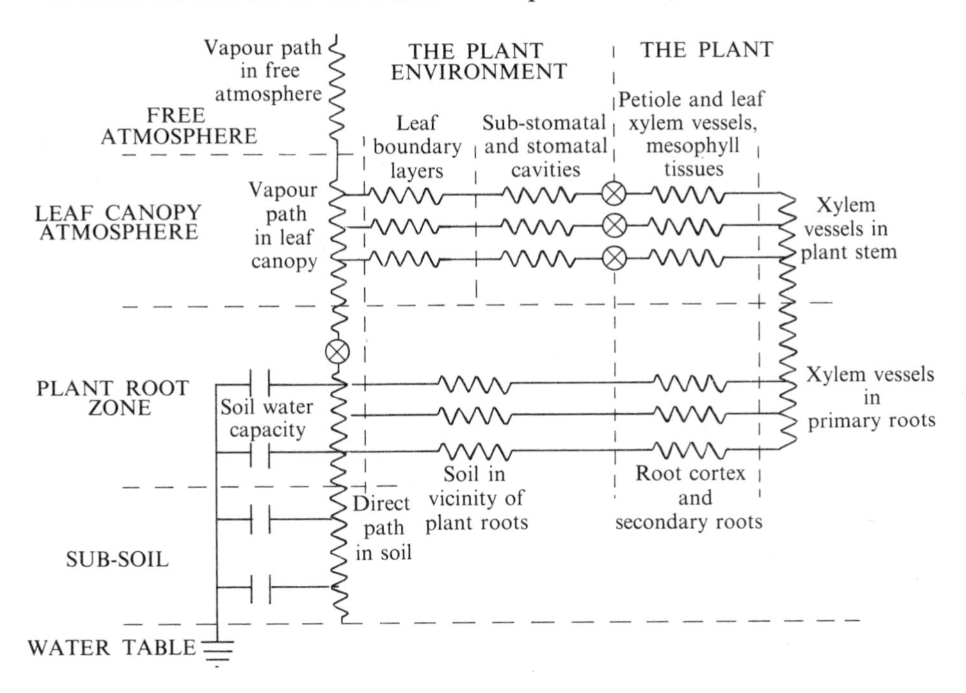

Fig. 3.1. Representation of pathways of water transport in the soil, plant and atmosphere. Sites of changes of phase from liquid to vapour are shown as ⊗. After Cowan, I. R. (1965). *J. appl. Ecol.* **2**, 221–39.

Table 3.1. Approximate magnitudes of water potential in the soil–plant–atmosphere continuum (J kg^{-1})

	Turgid plant	Wilting plant
Soil	−10 to −1000	−1000 to −2000
Leaves	−200 to −1500	−1500 to −3000
Atmosphere	−10000 to −200000	−10000 to −200000

amount used in expansion of new cells and 100 to 1000 times the amount used in photosynthesis. The system may be regarded as one in which water is supplied from the soil to the leaves to replace that being lost by evaporation and its components will be discussed in order along the pathway of flow. The treatment can only be approximate but, we hope, will draw attention to the main features. The system is depicted as an electrical analogue, using the convenient symbolism of electrical circuits, in Fig. 3.1. This may be regarded as representing a segment of a crop, well advanced in its growth and uniform over a large area; it will be referred to continuously in the following pages.

THE SUPPLY SYSTEM

Extent of the root system

Although roots of crops may penetrate 1–3 m in depth, the greatest concentration is in the upper layer (Table 3.2). It is convenient to express the concentration in length of root per unit volume of soil, R_V, and to consider its distribution in finite layers of the soil. In well-established crops, say at flowering, there may be 10–20 cm per cm^3 soil in the top 15 cm, this decreasing with depth to about 0.5 cm^{-2} in the 80–100 cm layer. Such a crop would have about 20 km roots per m^2 of ground area. Within 3–4 weeks of emergence of a densely sown wheat crop, roots may well have penetrated to a depth of 40–50 cm and have concentrations about one-tenth of those quoted above. After flowering, there is usually little further root extension and some decay of roots may occur during grain-filling. Most roots vary in diameter from about 5×10^{-2} (primary) to 10^{-2} cm (tertiary), have a surface area per unit length of 0.03 to 0.2 cm, and length per unit dry weight of 100–200 m g^{-1}. The development of the root system is discussed in Chapter 7.

Water absorption by roots

Consider the soil profile at some appreciable time after rain. The water content at different depths may then be somewhat as shown by the line *AO* in Fig. 3.2, the greater dryness in the upper layers reflecting the greater concentration of roots. Then assume that there is a heavy fall of rain or

Table 3.2. Root concentration, R_V (root length per unit volume), of established crops in the field. Reproduced from a compilation by Barley, K. P., *Adv. Agron.* (1970). **22**, 159–201, who cites original sources, plus data from Welbank, P. J. & Taylor, P. J. (1970). *Rep. Rothamsted Exp. Stn* p. 96, and Kirby, E. J. M. & Rackham, O. (1971). *J. appl. Ecol.* **8**, 919–24, assuming length/dry weight ratio of 150 m g^{-1}.

Species	Depth (cm)	R_V (cm^{-2})
Herbs: Gramineae		
Poa pratensis	0–15	50
Grasses	0–10	30–50
Cereals (oats, rye, wheat)	0–15	5–25
	25–50	4
	75–100	2
Wheat	0–15	8
	15–25	3
	25–55	1
	55–100	0.6
Barley	0–10	3.0
	10–20	1.2
	20–50	0.9
	50–100	0.5
	100–130	0.1
Herbs: non-Gramineae		
Stylosanthes gracilis	0–10	30
	40–50	3
	90–100	1
Medicago sativa	0–10	20
Glycine max	0–15	4
Woody plants		
Tea (*Camellia sinensis*)	0–2.5	4
	45–47.5	1
	68.5–70.0	0.5
Pinus radiata	0–8	2
	25–45	0.8
	91–106	0.4

that the soil is irrigated. Water enters the soil, and after a little time the moisture content through the profile is represented by *BLXO*. If the rate of entry is less than the rate of addition, water will be ponded on the surface and run-off may occur. The volume flux density of water, v – the volume of water crossing unit area per unit time – is proportional to the gradient of hydraulic potential, Φ, in the direction of the flux, i.e.

$$v = -k\,\mathrm{d}\Phi/\mathrm{d}z \text{ (cm s}^{-1}\text{)} \quad 3.1$$

where the hydraulic conductivity, k, varies widely with the pore geometry and the swelling characteristics of the soil. It is about 10^{-2}, 10^{-3} and

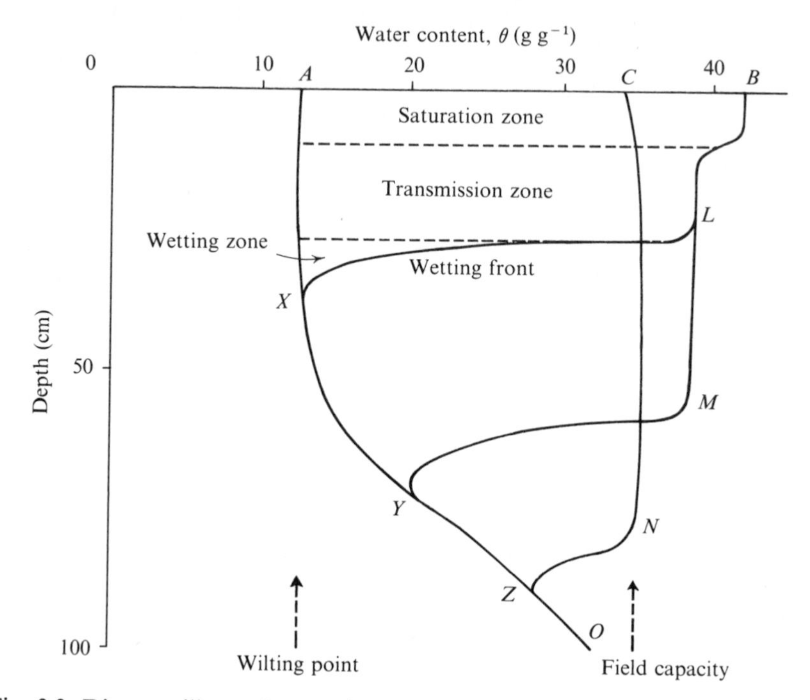

Fig. 3.2. Diagram illustrating wetting of a supposedly uniform soil with moisture characteristics similar in all layers of the top one metre. See text.

10^{-4} kg cm^2 s^{-1} J^{-1} in coarse sand, loam and clay soils, respectively. The difference between the moisture contents of the saturation and transmission zones reflects mainly the presence of entrapped air. Note that the wetting zone extends over a short distance only, indicating a sharp demarcation between completely wet and completely dry soil; this arises from the very low conductivity. The wetting front progresses downwards as long as the surface soil is saturated. We may suppose that *BMYO* represents the profile at the time that entry of water at the surface has ceased. After some further time, we find the profile *CNZO*, the continued movement representing drainage of the large pores; there would also be some loss by evaporation.

As long as there is excess water at the surface, the rate of infiltration (flux at the surface), v_s, approximates

$$v_s = 0.5bt^{-0.5} + a \qquad 3.2$$

where t is the time from the beginning of ponding. Both a and b vary with the pore space and arrangement and with the water content, the first representing the contribution arising from capillarity and the second that from gravity. As indicated, the flux falls rapidly with time and approaches a.

As a reasonable approximation we may take the following values of a (mm h^{-1}) for wet soils, deep sands and silts 8–11, sandy loam 4–6, many loams and clays 1–4, and high swelling soils 1. Approximate estimates may be obtained by taking these values after the first half-hour of ponding and using values up to five times greater than these during the first half-hour if the soil is initially very dry. If required the two constants can be readily evaluated experimentally. Estimates of the degree of ponding and hence of the probable amount of run-off then follow. If a consecutive budget of the soil water has been kept, the amount entering can be added; if this exceeds the soil water capacity of the root zone, then it may be assumed lost to the sub-soil and water-table. (More rigorous treatments are given in the references appended.)

As depicted in Fig. 3.1, absorption of water by the roots may be regarded as a flow in response to the decrease in water potential in the conducting vessels of the root xylem, this developing as a consequence of evaporation from the leaves. Regarding the uptake of water by a single cylindrical root of radius R as proceeding in a series of successive steady states, the constant rate, q per unit length, at any time is given by

$$q = -4\pi k(\psi_s - \psi_0)/\ln(A^2R^2) \text{ cm}^3 \text{ cm}^{-1} \text{ s}^{-1} \quad 3.3$$

where ψ_s and ψ_0 (here equal to τ_s and τ_0) are the water potentials in the soil at distance A and the root surface respectively. A can be taken to be about half the distance between neighbouring roots ($\simeq 1/(\pi R_v)^{0.5}$), and k is the capillary conductivity – analogous to the k for saturated flow stated in Equation 3.1 but varying over a range of several orders of magnitude as the soil dries (say, from about 10^{-4} or 10^{-5} around field capacity to 10^{-8} or 10^{-9} kg cm^2 s^{-1} J^{-1} near wilting). The capillary conductivity can often be described by

$$k = k_{sat}/\{(\tau/a)^n + 1\} \quad 3.4$$

where a is constant for any soil (about 1–10 J kg^{-1}) and n varies from about 2 in a clay soil to 10 in sand.

Similarly, the flow across the cortex of the root may be given by

$$q = -h(\psi_0 - \psi_x) \quad 3.5$$

where ψ_x is the water potential in the xylem (here, the osmotic potential must be included with τ because of flow across a semi-permeable membrane). Very few attempts have been made to measure h; one such measurement on young wheat roots indicated values of $0.2–2 \times 10^{-8}$ kg cm^2 s^{-1} J^{-1}. If this is typical, it suggests that roots provide the major resistance to flow in moist soils with the soil conductance becoming dominant as the soil dries. There should also be a factor incorporated in the above relations to take account of the increasing isolation of the larger roots from liquid water and the retreat of water into the smaller soil interstices as the soil dries.

On the other hand, some evidence suggests that a layer of mucilage surrounds roots, and this may provide a continued contact between roots and the soil water.

Water may also be expected to move from wetter soil below the root zone to replace that being lost. General experience shows that this is very small – because of the low conductivity. In most situations (unless the water table is less than 1 m below the root zone), it may be assumed that no water is so supplied.

It is much simpler, and possibly adequate for general purposes, to regard water uptake in the root zone as simply an upward flow per unit of soil surface. If v is the upward flux density at depth z, then

$$dv/dz = R_v(bk+h)(\psi_s - \psi_x) \text{ cm s}^{-1} \quad 3.6$$

where b is now a variable taking into account the degree of contact between the roots and soil water. This is permissible because the conductance for flow across the cortex appears to be 10^2–10^3 less than the conductance for flow along the xylem. Integration of Equation 3.6 over the several layers of the root zone then provides the total uptake.

Transport through the plant

The vessels from the various roots coalesce in the bundles of larger roots and finally all group together at the base of the stem. Bundles of xylem vessels exist within the stem and there are frequent anastomoses with petiolar bundles. The latter divide into those of the smaller veins and finally end as individual vessels adjacent to mesophyll cells in the leaf. At any point along this pathway, the flux is given by

$$v = f/A_x = -k_x d\Phi/dz \quad 3.7$$

where f is the total flow in the xylem of area A_x and Φ the hydraulic potential. The conductivity, k_x, varies with the number and diameter of the conducting elements in the functional xylem (i.e. by $\pi\Sigma_j\{n\}_j R_j^4/(8\eta)$ where n_j and R_j are the numbers and mean radii of the j size classes of *functional* vessels). k_x does not appear to vary greatly along the length of stems of herbaceous plants where the tissue is mature, being of the order of 2 kg cm^2 s^{-1} J^{-1}, but it may be only 10^{-2} of this value in young expanding stems and in petioles. This conductivity, together with the area of xylem present, is such that with the maximum rates of flow, the gradient in hydraulic and gravitational potential along the stems of both herbaceous plants and trees is unlikely to exceed about 20 J kg^{-1} m^{-1}.

There is much less certainty concerning the conductance in the leaf. Usually, it appears to be much less than that of the petiole. The lowest conductance may well reside in the pathway of flow of water from the vessel endings through the 2–3 cells before it evaporates. Most of this flow appears to occur along the cell walls although the potential of water here is likely to be fairly close to equilibrium with that in the protoplast.

WATER LOSS FROM A CANOPY OF LEAVES

Evaporation from a leaf

Water evaporates within the cell walls bordering on air spaces, diffuses through a short length of the cell wall, the sub-stomatal cavities, the stomata and the boundary layer over the leaf, and is then transported upwards by turbulent transfer. In all parts of the system, water moves from a region of high to one of lower concentration which may be expressed as mass of water vapour per unit mass of air, q. We can describe the mass flux of water vapour away from the leaf, E g cm^{-2} sec^{-1}, in the usual form, $E = -\rho K \, dq/dz$. But the system has various changes in geometry (with changing K) and the gradient cannot always be readily measured. The flux equation can be modified to the more convenient form

$$E = 2\rho(q'_s - q)/(r_a + r_l) \qquad 3.8$$

where q'_s is the saturation humidity at the evaporating surface within the leaf, q the humidity of the ambient air, and r_l and r_a are resistances of the pathway within the leaf and of the boundary layer, respectively, to flow of water vapour per unit of leaf surface. It is assumed that the resistances on the two sides are equal. (Essentially, $1/(r_a$ and $r_l)$ replaces K/l where l is the effective length between q'_s and q, but the resistance to flow beyond the boundary layer is relatively so small that it may be ignored.)

As q'_s is most difficult to measure in practice, we make use of another property. Imagine a mass of absolutely dry air in a perfectly insulated box. Its temperature T describes the content of *sensible heat* Tc_p, where c_p is the thermal capacity. Now introduce some liquid water at the same temperature. Water evaporates until the air is saturated; it then has a concentration q' and the temperature has been reduced. In other words, there has been a redistribution between sensible heat and *latent heat*, $\lambda q'$, where λ is the latent heat of evaporation; i.e. $\delta(\lambda q') = \delta(Tc_p)$. We can write $\delta(\lambda q')/\delta(Tc_p) = \epsilon$, where ϵ varies with the temperature.

Equation 3.8 can also be written in terms of the rate of loss of sensible heat, H, where

$$H = 2c_p\rho(T_s - T)/r_h \simeq 2\lambda\rho(q'_s - q)/\epsilon r_h \qquad 3.9$$

and q'_s and q' are now the saturation humidities at the surface temperature T_s and ambient temperature, respectively, and r_h is the boundary layer resistance to loss of heat per unit area of leaf surface. Combining Equations 3.8

and 3.9 and assuming that the net flux of radiation to the leaf, ϕ, is equal to the sum of the rates of loss of latent and sensible heat,

$$E = \{\epsilon r_h \phi/\lambda + 2\rho(q' - q)\}/(\epsilon r_h + r_a + r_l) \qquad 3.10$$

This equation holds only when the leaf resistances of the two surfaces are more or less the same. This is so with many crop plants under most situations. Where the resistances of the two leaf surfaces are widely different, r_a and r_l are replaced by $2R$, where $1/R = 1/(r_a + r_{lU}) + 1/(r_a + r_{lL})$ where r_{lU} and r_{lL} are the leaf resistances of the upper and lower surfaces, respectively.

In order to solve Equation 3.10, estimates are needed of ϕ, T, q, r_a and r_l. The quantities ϵ and q' are most readily obtained from tables of saturation vapour pressure, such as those given in the *Smithsonian Meteorological Tables*, taking $q = 6.22 \times 10^{-4} e$, where e is in mbar and $\epsilon = 1.515 \Delta e'/\Delta T$. r_h may be taken as $1.12 r_a$. With laminar flow, $r_a \simeq 2.6 L^{0.25} u^{-0.5}$, where L is the leaf area (cm^2) and u the wind speed (cm s^{-1}); under crop conditions, there is much uncertainty about the reliability of the value 2.6 and possibly a value around 1.7 may be more appropriate. A possible simplification is to take $r_a = 0.05/K_M$ (cf. Equation 3.16).

The resistance of each leaf surface, r_l, consists of the sum of the resistances due to the stomata, r_s, the sub-stomatal cavities, r_i, and the outer layers of the cell walls, r_w, in parallel with that of the cuticle, r_c. Inadequate estimates of the sub-stomatal-cavity and cell-wall resistances suggest that each may be about 0.2 s cm^{-1}. The minimum stomatal resistance rarely appears to be less than about 0.5 s cm^{-1}. The cuticular resistance of mesophytes (including crop plants) seems to be around 40–50 s cm^{-1} (range 20–80) whereas values ranging from 50 to 400 s cm^{-1} have been recorded for xerophytes. With crop plants, then, values of r_l ranging from a minimum of 1 s cm^{-1} might be expected. The diurnal curve of r_l can be measured directly using a so-called diffusion porometer. Alternatively, estimates of stomatal resistance can be obtained by converting measurements made using a viscous-flow porometer to diffusion resistances.

The stomata are the dominant variable in the whole of the vapour-phase pathway; indeed, they comprise the major varying *plant* component of the whole system. The three major variables to which the stomata respond are light flux, water and temperature. Some representative curves of the responses to these variables (Fig. 3.3) are described by the following functions:

Light flux:	$r_l = a + b/Q_v$	3.11
Leaf relative water content:	$\log r_l = a - b\zeta$	3.12
Temperature:	Uncertain	3.13

We still lack experience of the general validity of these functions and the range of, and factors influencing, the values of the constants. On general

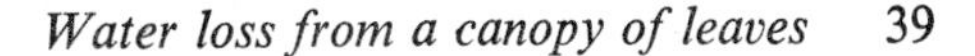

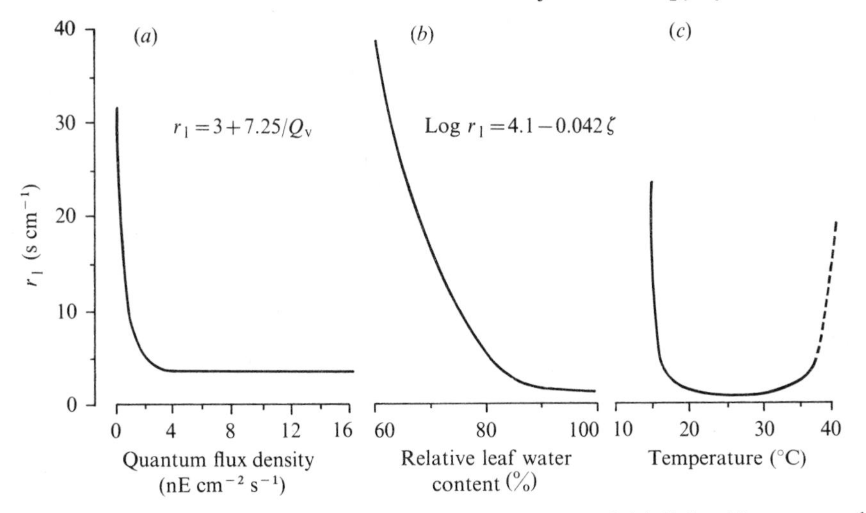

Fig. 3.3. Changes in stomatal resistance as a function of (*a*) light, (*b*) water and (*c*) temperature. Data in (*a*) from diffusion porometer measurements by Kanemasu, E. T. & Tanner, C. B. (1969). *Pl. Physiol., Lancaster* **44**, 1542–6; in (*b*) from chamber experiments by Troughton, J. H. (1969). *Aust. J. biol. Sci.* **22**, 289–302; and in (*c*) derived with several assumptions from silicone rubber impressions made by Hofstra, G. & Hesketh, J. D. (1969). *Can. J. Bot.* **47**, 1307–10 – this curve should approximate to the true form but may be appreciably displaced along the *y*-axis.

grounds, since $1/r_l = 1/r_c + 1/(r_s + r_i + r_w)$ (cf. above), we might expect r_l to vary between two asymptotes, a maximum representing r_c and a minimum representing the widest possible degree of opening of the stomata. That this does not clearly emerge from empirical measurements may be due, in respect of light, to the low flux densities to which the stomata respond. In the example cited in Fig. 3.3*a*, for example, the minimum r_l was obtained with a flux density equal to about 2 per cent full sunlight. In respect of water, the continually increasing resistance with decrease of leaf water (below incipient wilting) may reflect increasing cuticular and cell wall resistances, the latter possibly resulting from the retreat of the evaporating surfaces into the interstices of the cell walls. There is very little evidence on the effect of temperature, although some data suggest the relationship is of a sharp-sided U-form, with little change over a wide range of the usual 'physiological' temperatures and the temperatures at the lower and upper limits varying with species. The range over which r_s is approximately constant and minimal, for example, seems to be normally much wider than that illustrated in Fig. 5.9.

Evaporation from a crop

The calculation of evaporation from a canopy requires the determination of the net radiation, Q_N, available throughout the canopy. At any one point

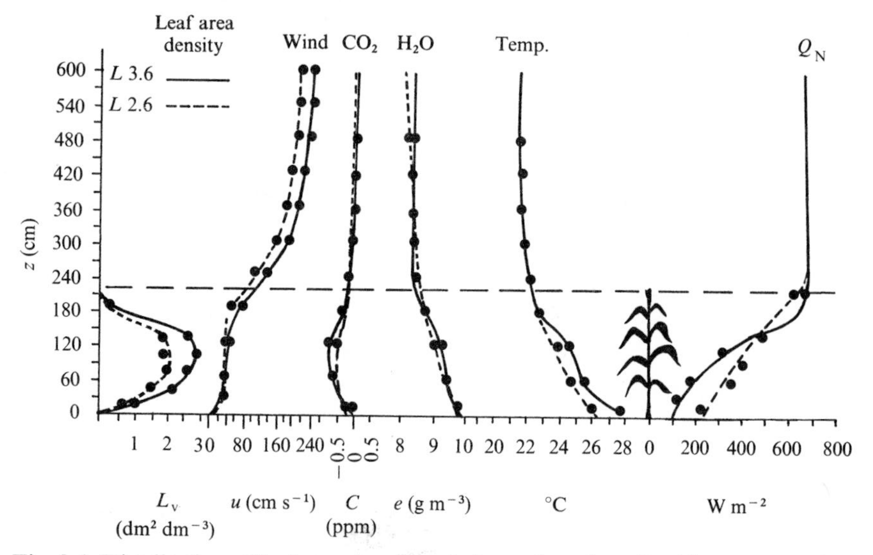

Fig. 3.4. Distribution of leaf area density, wind speed, carbon dioxide concentration, water vapour concentration, temperature and net radiation in maize crops. After Lemon, E. R. in Šetlík, I. (1969) (cf. Chapter 5).

Q_N is given by the difference between the downward flux of short-wave radiation (Equation 2.7) and the upward flux of long-wave radiation (Equation 2.8). The gradient at any point on the curve relating Q_N to z, the height above the soil surface, provides the amount of net radiation absorbed by the leaves at that point (Fig. 3.4); this is partitioned between the increases of latent and sensible heat as described on p. 37. Indeed Equation 3.9 can be written, for any thin layer of the canopy, as

$$dE/dz = \{\epsilon r_h(dQ_N/dz) + 2L_v \rho_a(q' - q)\}/(\epsilon r_h + r_a + r_l) \qquad 3.14$$

where E is the upward flux of water vapour and L_v the leaf area density. The fluxes of water vapour and heat, H, are also related to the gradients of humidity and temperature through the relations

$$E = -K_W \rho_a dq/dz; \quad H = -K_H \rho_a c_p dT/dz = -K_H(\lambda \rho_a/\epsilon)dq'/dz \qquad 3.15$$

The gradients indicate the direction of the flux but not its magnitude. (In Fig. 3.4, for example, the flux of Q_N is downward and those of E and H are upward throughout the entire canopy, whereas that of carbon dioxide is upward in the lower 60 cm and downward in the remainder of the canopy.) The magnitudes can only be determined if the appropriate transfer coefficient, K, is also known. There is still much uncertainty about the variation, and prediction, of K within a crop. One approximation is to take

$$K_H = K_W = \{3.0\text{–}1.4 \exp(1.5z/\chi)\}K_M \qquad 3.16$$

where K_M within the crop is given by

$$0.4u_* \chi\{1-\exp(-z/\chi)\}; \qquad 3.17$$

χ is a stability length defined as $\rho_a c_p T u_*^3/(0.4gH)$ and u_* is a function of the general crop structure and the wind speed at a reference height (cf. p. 28).

The fluxes and gradients which have been considered above are all intricately related. No simple statement of cause and consequence can be made without a reasonably complete analysis. Nevertheless, even allowing for the appreciable degree of uncertainty in some of the relationships presented above, it is possible to predict the water vapour and heat fluxes from a crop, and the distribution within a crop, knowing the variation of leaf area density with height, the resistances r_a and r_l, the incoming radiation, and the wind speed and humidity deficit at a reference height above the crop. The results of one such calculation (Fig. 3.5) illustrate the importance of the leaf resistance. In any given set of weather conditions, increase in leaf resistance leads to the transpiration rate becoming more uniform through the different layers, the total rate of loss decreasing and the temperature of the canopy increasing. However, the magnitude of these changes varies with the weather conditions outside the canopy.

This approach can also be extended to include estimates of water loss from the soil surface but there is great uncertainty concerning the changes in K in the layers just above the soil surface. Here, we provide a brief descriptive treatment only. During most of the daytime, there is a downward flux of net radiation at the soil surface which may be measured by soil-flux plates or estimated from radiation-interception curves. If the surface is wet most of the net radiation is used to evaporate water; evaporation of soil water may then account for up to one-third of the total flux at the crop surface. However, as the top few millimetres of the soil dry out, the resistance to diffusion of water vapour to the soil surface increases markedly. Some of the heat absorbed by the surface layers is then conducted downwards, some is transported upwards and a smaller and smaller proportion is used in evaporation. Possibly a total of some 15–25 mm – equivalent to the available water in the top 10–15 cm of soil – may be lost during any cycle of rewetting; thereafter, there is no further loss. The amount of water lost directly from the soil as a proportion of the total evaporation therefore depends mainly on the frequency of rewetting.

The rate of evaporation from the crop and soil as a whole can be obtained by measuring the gradients of wet- and dry-bulb temperatures over a finite height above the crop, thereby giving Δq and ΔT and hence the gradients of latent and sensible heat. The difference between the net radiation above the crop $Q_N(0)$ and at the soil surface $Q_N(S)$ gives an estimate of the amount of net radiation absorbed by the canopy. Then, using the relationships

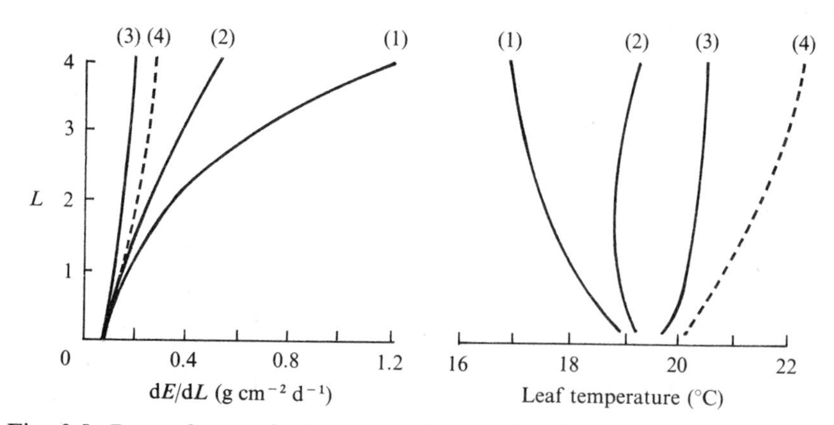

Fig. 3.5. Rate of transpiration per unit area (dE/dL) and leaf temperature as functions of the leaf resistance and the cumulative leaf area (L) measured from the soil surface. The data refer to a model canopy, of uniformly distributed leaves, receiving a total net radiation of 419 W m^{-2}. Curves (1), (2) and (3) refer to leaf resistances (r_l) of 0 (wet surfaces), 1.2 and 4.8 s cm^{-1} respectively, with values of 12 and 16 g m^{-3} for the water vapour content and saturated water vapour contents of the air at the top of the canopy. Curve (4) refers to a leaf resistance of 5.2 s cm^{-1} and water vapour contents of 11.1 and 18 g m^{-3} at the top of the canopy. After Cowan, I. R. (1968). *Quart. J. Roy. meteorol. Soc.* **94**, 523–44.

given by Equations 2.10 and 2.11 and assuming $K_W = K_H$ and $Q_N = E+H$,

$$E = \{Q_N(0) - Q_N(S)\}/(1 + c_p \Delta T/\Delta q) \qquad 3.18$$

Continuous measurements require elaborate instrumentation which often cannot be employed. Yet, estimates are frequently required from a paucity of information. Possibly the best procedure is to follow the analysis given above, accepting reasonable approximations where information is lacking. Accepting the dictum that the accuracy of the estimate depends on the accuracy of the information used in arriving at it, adequate estimates can often be obtained with very little data. A wide range of possibilities exists between the precision obtainable from applying Equation 3.18 or the canopy analysis outlined above and that from Equation 3.10 applied in its crudest variant to a whole stand. This equation was first derived by H. L. Penman in the form

$$E = f[\epsilon\{Q_N(0) - Q_N(S)\}/\lambda + E_a]/(\epsilon + 1) \qquad 3.19$$

where $E_a = 0.26(0.5 + 6.21 \times 10^{-3}u)(e' - e)$. Here, u denotes the wind speed at 2 m in km d^{-1}, $(e' - e)$ the vapour pressure deficit in mbar, and f varies from 0.6 to 0.8 depending on length of day and season. In this form, the relationship applies only to an extensive area of dense short vegetation well supplied with water, but it can be modified to take account of changes in leaf resistance as water becomes short. This relationship has proved enormously valuable and adaptable and it is an instructive exercise to trace its development from its original form. When first propounded in 1948, it provided a completely new approach to the understanding of crop–environment relationships in general and has been responsible, probably more than any other contribution, for the rapid progress that has been made in this field during the past two decades.

WATER BALANCE WITHIN THE CROP

Let us consider the changes in a crop following a heavy fall of rain when the water throughout the whole root zone has been replenished to field capacity. The leaf cells are near to full turgor (Fig. 3.6). Both diurnal and longer-term changes occur. Soon after dawn, light falling on the leaves leads to stomatal opening and the incoming radiation provides energy for evaporation. Water is lost from the cell walls and flows from the included protoplasts and from adjacent cells to maintain equilibrium of water potential. The loss of water from the cell leads to a decline in volume and turgor pressure P, and hence ψ. The flow from adjacent cells leads to the establishment of a gradient of water potential back through the veins and the stem to the root and to the soil surrounding the root. The sequence of changes which occur may be illustrated using a simplified model in which the water potential in the bulk soil, at the root surfaces and in the leaves is followed (Fig. 3.7). As the radiation and rate of evaporation increases during the day (e.g. Day 1), the water potential in the leaves decreases, reflecting the greater potential difference needed to maintain higher flow rates with a constant resistance in the pathway. With the decreasing rates of evaporation in the afternoon the potential difference between leaves and soil also decreases, until finally at night equilibrium is again established – at a water potential which effectively is that of the water in the soil. On Day 2, the same process is repeated but here it is assumed that the leaf water potential reaches a critical value (-1500 J kg^{-1}) at which the stomata close completely. With the decreased rate of water loss, despite the high radiation, the leaf water potential increases. And so the diurnal cycle is repeated with the soil becoming drier and its water potential decreasing, and the leaf water potential reaching lower values for successively longer periods each day.

The position and, indeed, the form of the curves shown in Fig. 3.7 vary, as discussed above, with a number of parameters, particularly the hydraulic properties of the soil, the concentration of roots in the different layers of the soil, the weather conditions influencing evaporation, the stomatal responses, and the internal resistances to flow within the crop. In particular, attention is drawn to marked variation in water content and, hence, water potential in the different layers of the soil as it is dried out (Fig. 3.8). This mainly reflects the distribution of root concentration in the soil, most water

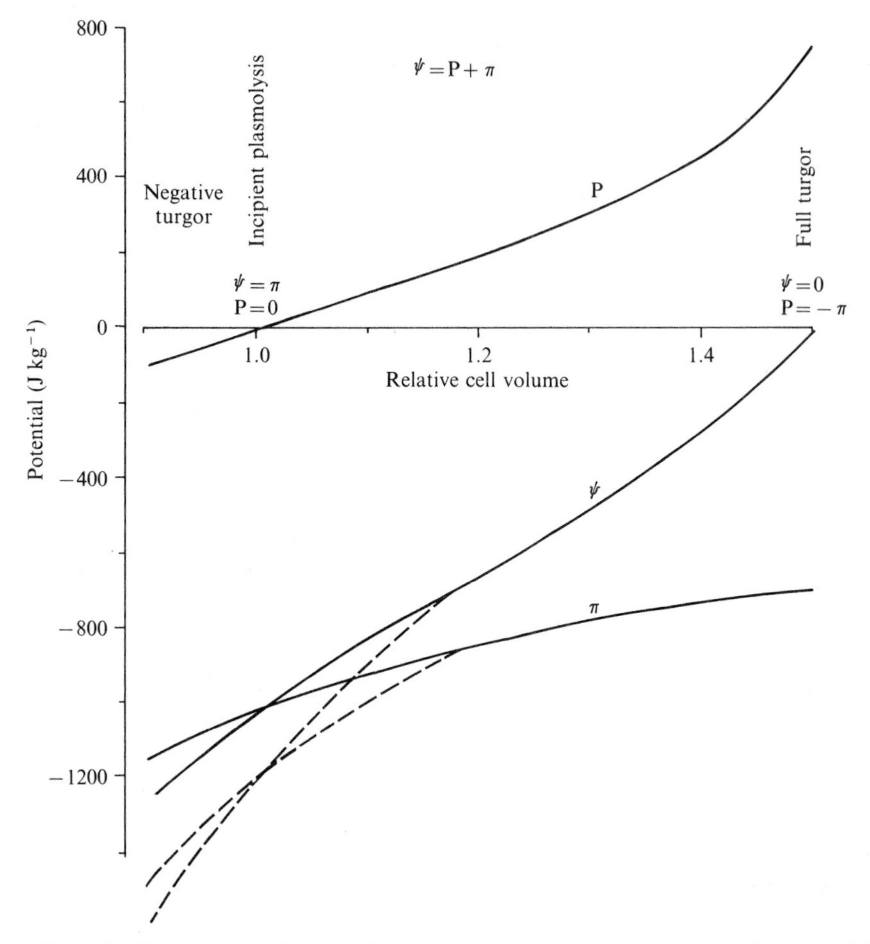

Fig. 3.6. Diagram showing relationships of turgor pressure, P, osmotic potential, π, and water potential, ψ, over a range of cell volumes from incipient plasmolysis to that at full turgor. The continuous lines for ψ and π refer to the situation where there is no increase in content of osmotically active substances, and the broken lines where these substances increase with decrease in turgor.

being lost from the regions with the greatest concentration of roots. There may also be a marked gradient of water potential in the different layers of the canopy, but too few measurements have yet been made to generalize. Gradients at the same point of time from the upper to lower leaves of −1000 to −400 in tobacco, −1800 to −2100 in bulrush millet and −450 to −150 J kg^{-1} in maize have been recorded. On the other hand, the relative water content (i.e. that held at any time relative to the amount at full turgor) of lower leaves always seems to be less than that of upper leaves despite

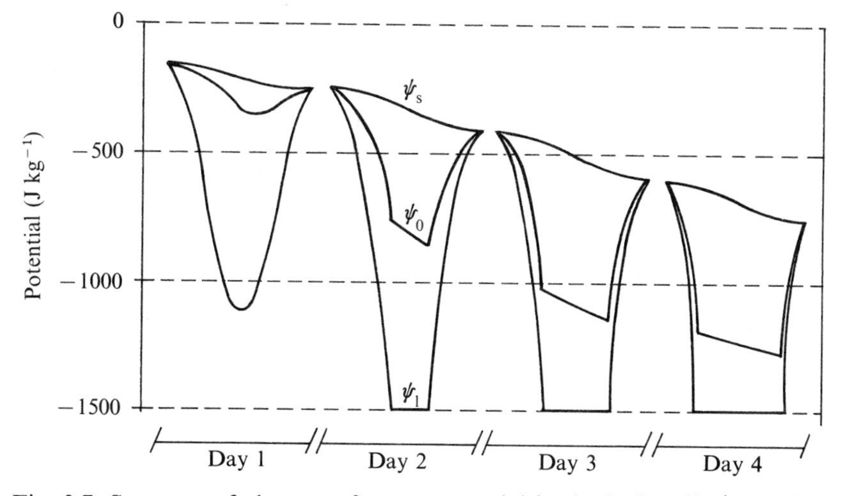

Fig. 3.7. Sequence of changes of water potential in the bulk soil, ψ_s, at the root surfaces, ψ_o, and in the leaf, ψ_l, during a drying period. After Cowan, I. R. (1965). *J. appl. Ecol.* **2**, 221–39.

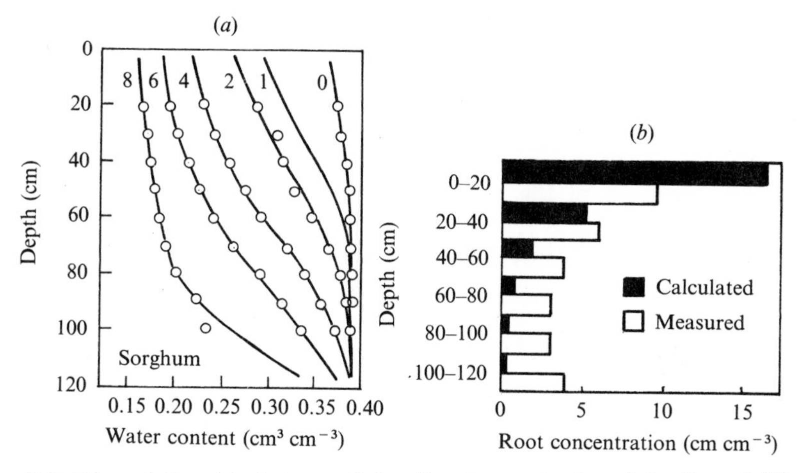

Fig. 3.8. The relationship between (*a*) soil-water content and depth and (*b*) the root concentration and depth during a drying cycle. Curves 0–8 represent the respective numbers of days after wetting. The circles indicate experimental data, and the curves were calculated from theory. Similarly, the white histograms were actual measurements and black histograms calculated amounts. After Gardner, W. R. (1964). *Agron. J.* **56**, 41–5.

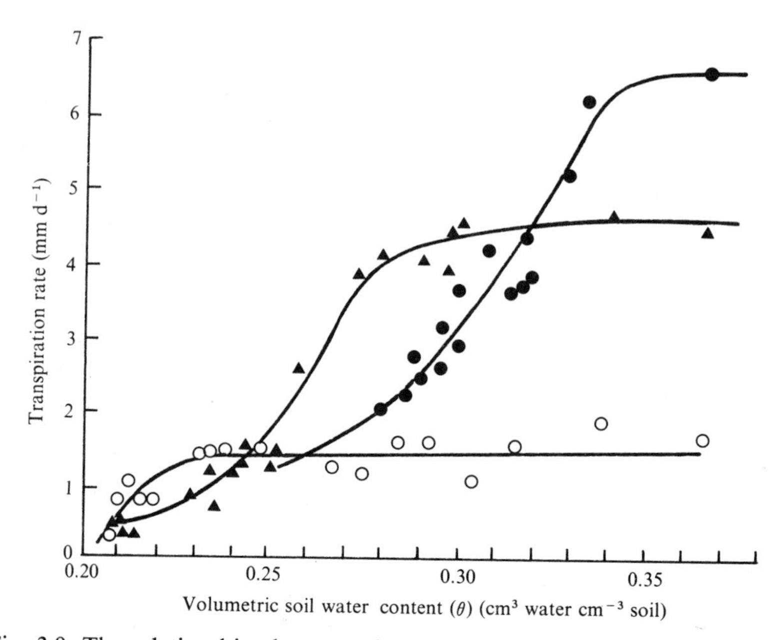

Fig. 3.9. The relationships between the actual transpiration rate of a maize crop and soil water content on days of high (closed circles), medium (triangles) and low (open circles) potential transpiration. The potential transpiration rate is indicated by the values at the highest soil water content. After Denmead, O. T. & Shaw, R. H. (1962). *Agron. J.* **54**, 385–90.

the lower resistance of the petioles to flow and the lower net radiation received.

Fairly close interrelationships exist between the actual transpiration rate, the potential transpiration (i.e. the transpiration with full stomatal opening and hence determined only by the weather) and soil water content (Fig. 3.9). When the potential for transpiration is high, it can only be met by the plant when the soil is wet; as the soil dries, water cannot flow into and through the plant quickly enough to meet the demand, the leaf water potential falls, the stomata close partially and the transpiration rate decreases. Curves such as these can be determined empirically or, better, constructed from known values of the determining parameters, and used for forecasting changes in transpiration (cf. Kozlowski, 1968). In such situations, it is usual to make daily assessments of the soil water balance and the plant water status, the latter being measured either as the water potential or as relative water content of the youngest fully expanded leaf to which growth rate and other physiological processes are related. These responses are discussed later.

FURTHER READING

Eckardt, F. E. (ed.) (1968). *Functioning of terrestrial ecosystems at the primary production level.* UNESCO, Paris.

Fogg, G. E. (ed.) (1965). *The state and movement of water in living organisms. Symp. Soc. exp. Biol.* vol. 19. Cambridge University Press.

Kozlowski, T. T. (ed.) (1968). *Water deficits and plant growth,* 2 vols. Academic Press, New York and London.

Kramer, P. J. (1969). *Plant and soil water relationships. A modern synthesis.* McGraw-Hill, New York.

Penman, H. L. (1963). *Vegetation and hydrology,* Tech. Comm. No. 53, Commonwealth Agricultural Bureau, Farnham Royal.

Slatyer, R. O. (1967). *Plant-water relationships.* Academic Press, New York and London.

4

The Absorption and Transport of Mineral Nutrients

THE AVAILABILITY OF IONS IN THE SOIL

The major source of mineral nutrients is the soil and, although seeds contain considerable quantities of mineral nutrients, uptake from the soil starts soon after germination and soon contributes most of the ions entering the root and shoot axes.

Ions occur in the soil in a variety of forms. Some are adsorbed on the soil surface and others are associated with the various charged sites in the crystalline structure of the clays, and in the soil solution. There is continual exchange between these various phases, and if extra ions are added as fertilizers they are distributed between the phases (cf. Chapter 2) and few remain in the soil solution (Table 4.1). There is only a very poor correlation between the growth made by a plant and the total amount of any particular ion in the soil. The reason for this is that not all the ions are in a form that can be absorbed by plants, i.e. only a proportion of the ions are 'available'. The available ions are those present in the soil solution and those in the solid phase which are easily exchangeable with the soil solution; these form the so-called labile pool of ions.

Various means have been used for measuring the amount of available ions and most suffer from some defects. The absorption of ions from soil is a complex process and is difficult to simulate in a single test procedure made at one sampling time. It is for this reason that simple extractions with various solvents are not always very successful. A further complication is that where an extractant can be used to estimate the exchangeable ions (e.g. ammonium acetate or chloride for potassium ions) these are not always all equally available to plants.

The principles, and problems, associated with estimates of availability can be illustrated by reference to the methods used to estimate the amount of available phosphorus. We need to know the amount immediately available, i.e. the concentration of phosphate in the soil solution, the amount of 'labile-P' which can exchange with that in solution, and the rate at which this exchange can take place. The concentration of phosphate in solution is directly measurable, but the estimate obtained is dependent on the ratio of

Table 4.1. Concentration of phosphorus in solution applied, and after coming to equilibrium with three Australian soils (both in ppm; phosphorus applied in 0.01 M calcium chloride solution). After Biddiscombe, E. F., Ozanne, P. G., Barrow, N. J. & Keay, J. (1969). *Aust. J. agric. Res.* **20**, 1023–33.

Crawley sand		Coolup sand		Dardanup loam	
Applied	Equilibrium	Applied	Equilibrium	Applied	Equilibrium
0	0.016	0	0.002	0	0.001
5	0.067	12.5	0.002	20	0.003
10	0.24	25	0.02	50	0.004
25	*	50	0.043	100	0.005
50	*	100	0.259	1000	0.075

* This soil has little ability to adsorb phosphate and appeared to be almost saturated by these additions of phosphate. Hence the buffering capacity for phosphate of these soils to which phosphate has been applied approaches zero, and it is not possible to interpolate a phosphorus concentration which would be in equilibrium with the soil.

soil to solution, the concentration of ions in the extracting solution and several other factors. The method has, therefore, to be standardized; for example, by extracting one part of soil with ten parts of 0.01 M $CaCl_2$ solution at a known temperature for a known time and then estimating the amount of phosphate in the solution. The second and third needs can be reasonably satisfied by a single test in which the amount of phosphorus which can come into solution within a reasonable time span is measured. This is most readily achieved by isotopic dilution, making use of the relation

$$\frac{^{32}\text{P on soil surface}}{^{31}\text{P on soil surface}} = \frac{^{32}\text{P in solution}}{^{31}\text{P in solution}} = \frac{^{32}\text{P in plant}}{^{31}\text{P in plant}} \qquad 4.1$$

Two methods are used: one, 'the E-value' (after 'exchangeable'), being a short-term laboratory technique, and the other, the L-value (after S. Larsen), which necessitates growing plants over a period of several weeks. In determining the E-value the soil is shaken with a solution of ^{32}P of known *high* specific activity for a given time, t (about a week). (It is necessary to use as high a specific activity as possible to avoid affecting the exchange rates by the addition of large amounts of phosphate.) At the end of the equilibration period the soil is separated from the solution by centrifugation and filtration and the amount of ^{32}P and stable ^{31}P remaining in the solution determined. If x and x_t are the amounts of ^{31}P present in the solution initially and after

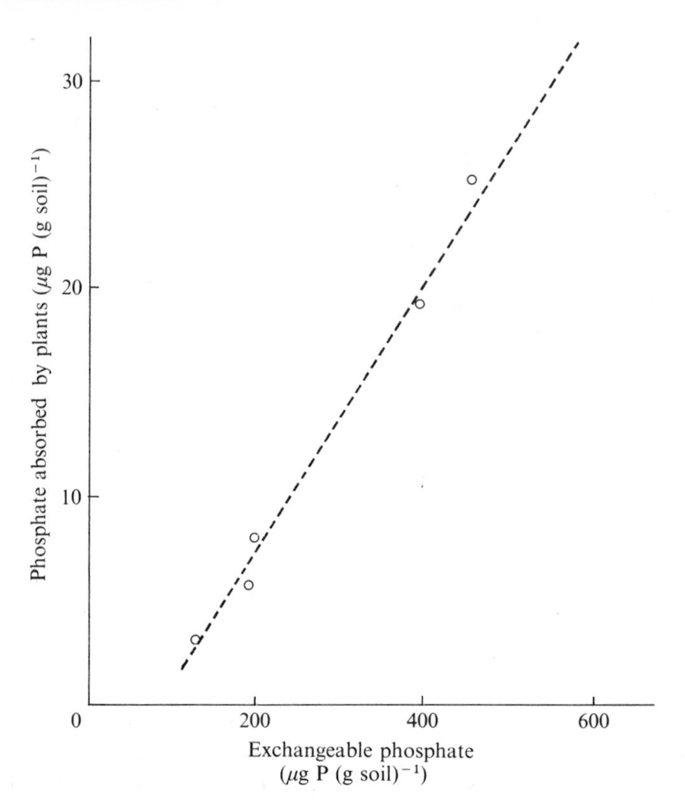

Fig. 4.1. The relationship between the absorption of phosphate by barley plants in 43 days and the isotopically exchangeable phosphate, E, in a medium heavy loam treated with varying quantities of soluble phosphate. After Russell, R. S., Barber, D. A. & Newbould, P. (1961). *Outlook on Agriculture* **3**, 103–10.

time t, and y and y_t are the analogous amounts of ^{32}P, then if the exchange of ^{32}P has come to equilibrium the specific activity is given by

$$y_t/x_t = y/(x+\mathrm{E}) \qquad 4.2$$

Therefore

$$\mathrm{E} = \{(yx_t)/y_t\}-x \qquad 4.3$$

The estimate of E obtained is a function of the ratio of soil to solution, the temperature, the period of shaking and, over short periods, the amount of ^{31}P in the original solution. In addition to the exchange processes, some of the added phosphate will be lost from the solution by sorption, S; this will form part of the exchangeable phosphorus and can be estimated from $S = x-x_t$.

The uptake of ions can be considered in a manner analogous to water absorption, i.e. in terms of a movement from a region of a higher to a lower

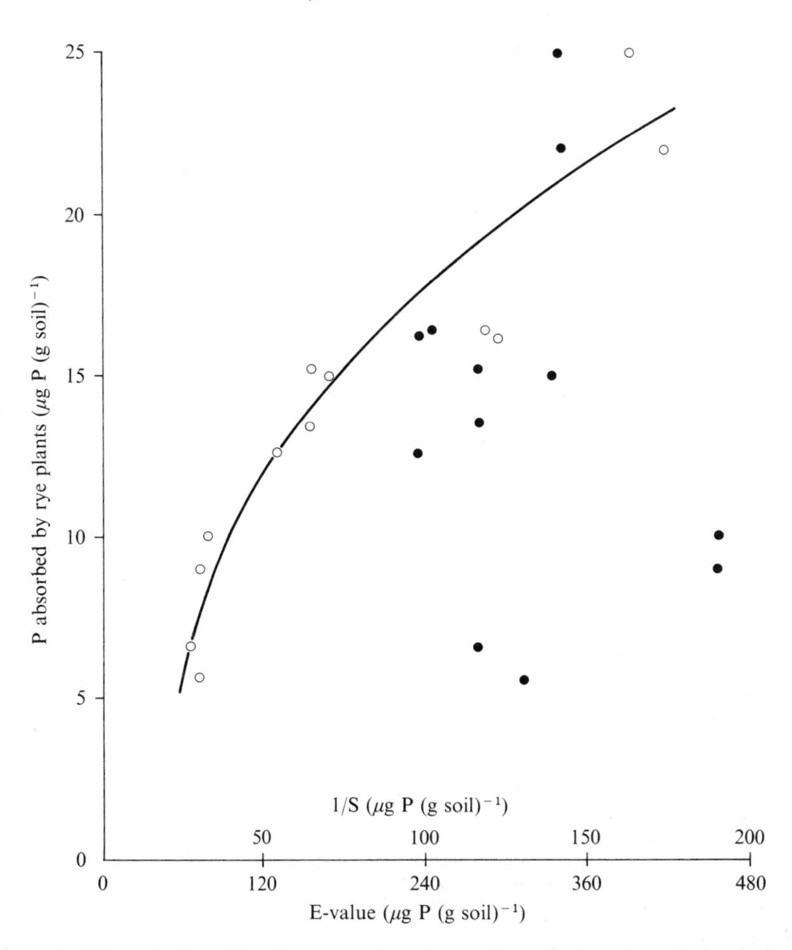

Fig. 4.2. The relationships between the absorption of phosphate by rye plants in 65 days and the isotopically exchangeable phosphate, E (closed circles) and reciprocal of the sorbed phosphate, 1/S (open circles and curve). Three contrasting soils were used, each enriched to two levels of phosphate and stored for 118 days at either field capacity or air-dry. Redrawn from Russell, R. S., Russell, E. W. & Marais, P. G. (1957). *J. Soil Sci.* **8**, 248–67.

potential. As absorption proceeds, the ionic potential of the solution decreases and more ions are released from exchange sites. This situation has led to the suggestion that another factor should be considered: this is the buffering capacity or the ability of the soil to maintain a given potential as the ions are removed by plants. It is given by the slope of the line relating the concentration in the soil solution to the amount of phosphate in the labile pool.

Table 4.2. Percentage of variation in phosphate uptake by *Lolium perenne* accounted for by selected soil parameters. After Gunary, D. & Sutton, C. D. (1967). *J. Soil Sci.* **18**, 167–73.

	Initial P uptake (first cut of ryegrass)			Total P uptake (sum of six cuts of ryegrass)		
Amount of phosphate added*	0	560	1120	0	560	1120
log [P]	49.4	41.6	19.0	59.2	42.4	36.9
L-value	56.5	21.2	2.0	55.7	13.6	11.1
L-value + log [P]	73.5	87.8	29.3	79.6	81.2	63.2
Resin P	69.9	48.1	18.4	69.4	59.0	59.6

* kg P_2O_5 applied per ha five years previously.

The use of only one of the three factors – soil solution concentration, labile phosphate and the buffering capacity – cannot give a complete description of the overall system and therefore may, or may not, provide a successful estimate of the amount of phosphate available to plants. For example, in experiments where varying amounts of fertilizer are added to one soil there will be a single relationship between the three factors and the use of only one, the E-value, is often sufficient (Fig. 4.1). With contrasting soils, however, each with their own relationship, the use of the E-value is often not adequate but there can be a relationship with the reciprocal of the sorption (Fig. 4.2). The phosphate potential in these soils would be a function of the concentration of phosphate in solution and hence inversely related to that lost from solution, i.e. sorbed.

If two factors are used to characterize the available phosphate instead of only one, it should be possible to obtain a much better relationship between uptake and soil phosphate. A direct attempt to relate phosphate uptake by *Bromus mollis* to the phosphate potential and buffering capacity was made, using 42 different Australian soils, and yielded the regression

$$\text{P uptake} = 0.0375\,(t\text{-}21)(j)(8.92+0.0043t+\text{potential})^2 \qquad 4.4$$

where uptake was in mg P per pot, t was time from sowing in days, j was a measure of buffering capacity and the potential measured was

$$-(0.5\log[Ca^{++}]+\log[H_2PO_4^{-}])$$

i.e. the negative potential of $Ca(H_2PO_4)_2$. This regression accounted for about 90 per cent of the variation in phosphate uptake.

A more indirect attempt to explore this relationship used the logarithm of the phosphate concentration and the L-value as an estimate of the labile phosphate (Table 4.2). Neither of these factors accounted for much of the phosphate uptake when considered alone, but when combined in a multiple

Table 4.3. Percentage of the sums of squares for cumulative uptake of sulphur accounted for by the sources of variation shown in an experiment which measured the absorption of sulphate by *Lolium rigidum* from soil differing in buffering capacity due to the amount of sulphate in the soil solution. The number of significant regression terms is given in parentheses for each value. After Barrow, N. J. (1969). *Soil Sci.* **108**, 193–201.

	Harvests (weeks after planting)					
Source of variation	7	10	13	17	20	20+ roots*
Full regression	72.6 (6)	83.2 (6)	87.8 (5)	88.2 (4)	87.5 (4)	87.3 (4)
Concentration of sulphate in soil solution	38.8 (3)	57.8 (3)	67.5 (3)	70.0 (2)	72.4 (2)	75.1 (2)
Estimate of buffering capacity	33.8	25.4	20.3	18.2	15.1	12.1

* The first four harvests were of shoots and leaves. At the fifth harvest the roots were also harvested.

regression accounted for most of the variation. At low phosphate concentrations the estimates of solution and labile phosphate appeared to be equally important, and this suggests that the plants may have been rapidly exhausting the available phosphate near the roots. The decreasing proportion of the phosphate uptake that can be accounted for by all measurements at the higher phosphate concentrations is probably indicative of the increasing importance of other factors in limiting growth. The relatively faster decrease in importance of the L-value, however, suggests that at high phosphate concentrations the plants deplete the phosphate less rapidly and hence the amount of labile phosphate becomes less important. Greater precision could probably be obtained by taking account of the changes in the rates of ion uptake and growth throughout the experiment, and especially of variations in the rates of growth of roots and their exploration of the soil. An easier technique than measuring phosphate concentration and L-values is to measure the transfer of phosphate from the soil to an ion-exchange resin. This technique estimates both the phosphate in solution and that exchanging from the soil phase of the soil during the period of equilibration, and so gives a composite measurement of both the buffering capacity and potential.

Similar considerations to those described above apply to the uptake of sulphate, and a multiple regression combining the concentration of the sulphate in the soil solution and an estimate of buffering capacity has been found to be more effective than the use of either of the variables alone (Table 4.3).

Another factor influencing the ability of plants to absorb phosphate, or any other ion, is the rate at which the roots grow into unexploited regions of the soil or proliferate in one region, for example near fertilizer. The pattern shown in Table 4.2 in the changes in the relative importance of the two factors over long and short time intervals suggests that continued uptake may lead to the successive depletion of phosphate from limited regions of soil as the roots extend (cf. Fig. 4.5). Other evidence indicates that uptake from any one region of the soil can be considered to have ceased about five days after it was penetrated by a root. Alternatively, most of the phosphate entering a plant does so through tissues less than six days old (see below).

The amount of available P in the soil decreases as uptake proceeds throughout the growing season but is increased following rain, irrigation or the application of fertilizers. Only a fraction of the phosphate applied as fertilizer is absorbed in the year of application. The partitioning between the labile and non-labile pools of ions depends on the anion exchange sites in the soil and varies with soil type. Those soils in which the phosphate is strongly adsorbed are termed 'phosphate fixing', and plants growing on soils of this type would absorb in the first year much less than the 10 per cent of applied phosphate mentioned in Chapter 2.

Similar considerations to those outlined above apply to the measurement of the availability of ions such as potassium and calcium, although the uptake of other ions has not been studied as extensively as that of phosphate. When, for example, plants have an ample supply of water, uptake of potassium can be correlated with estimates of 'exchangeable' potassium, but this correlation breaks down when water is in short supply. The determination of E-values for potassium is hindered by the lack of a suitable radioactive isotope of potassium: ^{40}K has a half-life of only 12 hours and, therefore, ^{86}Rb or ^{137}Cs has sometimes been used in its place. Similarly, difficulties in measuring the low-energy β-emission from ^{45}Ca have led to the use of ^{89}Sr or ^{85}Sr. These can lead to further problems, however, because of discrimination by the plants between K, Rb and Cs, and Ca and Sr.

The issue of measuring the availability of nitrogen is complicated by the presence in the soil of several different forms which can be absorbed by plants (cf. Fig. 2.2). A satisfactory estimate would have to take account of the relative proportions of these substances in the soil, their relative rates of absorption by plants and the rates of interconversion between them. The processes of nitrification and denitrification and the immobilization and volatilization of ammonium ions are all complex functions of the soil temperature and water potential, oxygen supply, and sometimes pH. Some indication of the values of the parameters concerned, including changes throughout the growing season, are given on p. 12 but there are, as yet, insufficient data available to be able to provide a quantitative description of these relationships. The high solubility of ions such as NH_4^+ and NO_3^- means that it is necessary to account for losses caused by leaching.

At present, a number of inadequate methods are used to estimate the amount of available nitrogen. One that is often used is to measure the total nitrogen content of the soil by a Kjeldahl method. This accounts for only about 15 per cent of the variation in nitrogen uptake and plant growth in a variety of crops in New South Wales. Another method which is sometimes useful is to measure the amount of nitrogen which is extracted by hot water under standard conditions.

ABSORPTION OF IONS BY ROOTS

The ions absorbed from the soil by a root arrive at the root surface because they have moved through the soil or the root has grown into a previously unexploited region of soil. Which of these processes is the more important at any one time depends on the ion in question, the soil type and the plant involved. However, a root does not necessarily absorb all the ions with which it comes in contact; this discrimination between ions, or differential permeability, will be discussed later (p. 64).

Even if there is little or no root growth, the soil solution is likely to be moving because of the percolation of water after rain or its flow to replenish that lost by transpiration and evaporation. Ions present in the soil solution will, therefore, be transported to the root surface by this mass flow with a flux

$$F_f = vC \qquad 4.5$$

where v is the apparent velocity of water movement to the root. Depending on the relative rates of absorption by the roots and convective movement in the soil solution, the concentration of ions at the surface of the roots can change from that in the bulk solution. A diffusion gradient will, therefore, be established and the ion will tend to diffuse towards or away from the root depending on the direction of the gradient, the flux (F_d) being given by

$$F_d = D\xi \mathrm{d}C/\mathrm{d}z \qquad 4.6$$

where D is the diffusion coefficient of the ion in the soil solution, ξ is a dimensionless factor used to account for the diffusion in a cylindrical system, and $\mathrm{d}C/\mathrm{d}z$ is the concentration gradient of the ion between C_r and C – the concentrations of the ion at the root surface and in the bulk soil solution respectively. ξ is a function of Dt/R^2 where t is the time from the start of uptake and R the radius of the root; it decreases with time and over a large range of values of Dt/R^2 approximates to 1. This diffusional flux is also dependent on the soil structure, the water content of the soil and the presence of other ions.

Table 4.4. Estimates of the contribution of different types of root member of 4-week-old barley (cv. Maris Badger) to the uptake of phosphate and calcium. Seminal roots are those present in the embryo, nodal roots are initiated after germination. After Russell, R. S. & Newbould, P. in Whittington (1969).

	Seminal axes	Nodal axes	Laterals
Total length (m)	4.2	3.8	48
Percentage contribution to total uptake of:			
Phosphate	10	30	60
Calcium	10	45	45

The overall flux of an ion to the root, F, may therefore be written

$$F = F_d + F_f = D\xi \, dC/dz + vC \qquad 4.7$$

No account has been taken here of the growth of the root into unexploited soil. The treatment of this is still speculative, but under conditions, for example, of low values of v and D and the rapid depletion of ions from the regions close to the root, the rate of ion uptake can be a function of the rate of root growth.

If ion uptake is considered simply as the transfer of ions across the root surface it can be written as

$$F = 2\pi\bar{R}\bar{\alpha}LC_r \qquad 4.8$$

where $\bar{R}$ (cm) is the mean radius of the roots, L (cm) their total length and $\bar{\alpha}$ a transfer coefficient averaged over the whole root surface and which estimates the efficiency of ion uptake by the roots. It has been known for a long time that the rate of ion uptake per unit area of root surface changes with time and along the length of a root axis; there are also differences between the various axes within a root system. A more complete analysis would therefore take account of these changes. Although there are as yet insufficient data to allow us to generalize about such effects, the results of one such analysis are shown in Table 4.4.

The analysis described by Equations 4.5 to 4.8 is useful in that it helps to distinguish between the various types of movement to roots but it is difficult to use experimentally for this purpose; nor is it easy to obtain an estimate of $\bar{\alpha}$. One way in which the analysis can be extended is to consider the rate of ion uptake by the plant.

If the total weight of a plant is W and the mean concentration of an ion in the plant is X, then the flux of the ion into the plant is given by

$$F = d(WX)/dt = \{(dW/dt)(X) + (dX/dt)(W)\}$$

$$= WX\{(dW/dt \, . \, 1/W) + (dX/dt \, . \, 1/X)\} \qquad 4.9$$

Table 4.5. The change with time in values of $\bar{\alpha}$, the mean transfer coefficient for phosphate into the roots, of three species grown at a range of phosphate concentrations. Unpublished data of E. K. Christie.

Species	External concen- tration (ppm)	Values of $\bar{\alpha}$ (μm s^{-1}) at Day no.: 4	7.5	11	14.5	18	21.5	25
Thyridolepsis mitchelliana	0.003	−1545	45	226	230	238	291	440
	0.03	876	350	194	152	170	273	630
	0.3	284	155	110	102	124	195	399
	3	33	19	14	11	12	14	20
	30	4	3	2	2	2	1	1
Astrebla elymoides	0.003	3213	1346	655	368	237	177	151
	0.03	727	369	233	188	197	271	490
	0.3	315	162	102	77	71	79	107
	3	29	15	9	7	6	7	10
	30	3	1	1	0.8	0.9	1	2
Cenchrus ciliaris	0.003	1527	545	253	153	121	127	175
	0.03	821	404	247	189	180	214	317
	0.3	216	156	119	95	80	70	64
	3	23	20	19	20	22	27	37
	30	2	3	3	3	4	4	5

and by combining Equations 4.8 and 4.9 we get

$$2\pi\bar{R}\bar{\alpha} = (WX/LC_r)\{(dW/dt.1/W)+(dX/dt.1/X)\} \qquad 4.10$$

where $\bar{R}\bar{\alpha}$ is an estimate of the mean efficiency of the root system to absorb the ion and can be obtained from simple measurements of plant parameters. It can be seen that it is a function of the ratio of plant dry weight per unit length of root (W/L), the ratio of the concentration of the ion in the plant to that at the root surface (X/C_r), the relative growth rate ($dW/dt.1/W$) and the relative rate of change in the concentration ($dX/dt.1/X$). The most difficult parameter to measure is C_r. With well-stirred nutrient solutions it is usually sufficient to assume that $C_r = C$, the bulk concentration. Some of the results from an experiment of this type are shown in Table 4.5. The three grass species are found in soils of widely differing phosphate concentrations and the optimum phosphate concentrations in solution were 0.003, 0.03 and 0.3 ppm for *Thyridolepsis*, *Astrebla* and *Cenchrus* respectively. As might be expected, $\bar{\alpha}$ declined as the concentration of phosphate in the external solution increased. The change in $\bar{\alpha}$ with time was more complicated, there being a decrease up to about fourteen days and a subsequent increase, but with the minimum value being affected by the external concentration. It is apparent that optimum concentrations for growth could not be ascribed

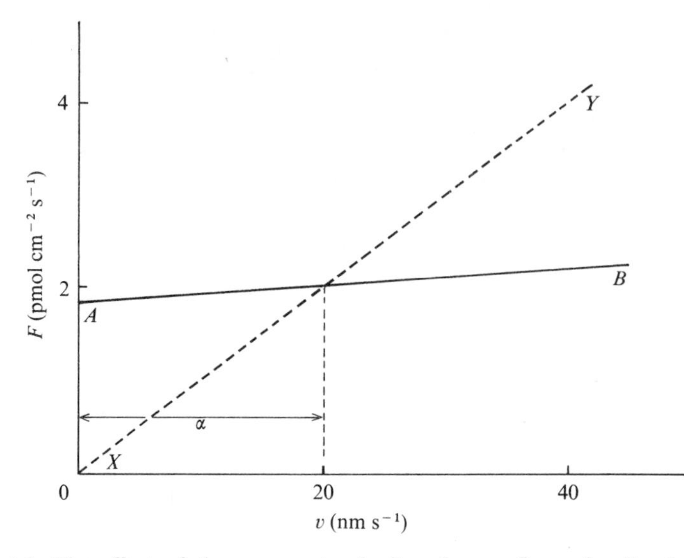

Fig. 4.3. The effect of the apparent velocity of mass flow of soil solution to root surface, v, on the flux of an ion into the root, F. Line AB assumes no diffusional component to the total flux and line XY the flux when diffusion does play a part. At $v = \alpha$ diffusion has no effect on F; when $v < \alpha$, F is reduced, and when $v > \alpha$, F is increased compared with the uptake in the absence of diffusion. The values given approximate to those for magnesium. After Tinker, P. B. in Rorison (1969).

simply to differences in $\bar{\alpha}$. The ability of the plants to use the phosphate was also important and there were differences here also, with, for example, *Thyridolepsis* transporting relatively more of the absorbed phosphate to the aerial parts than did the other two species.

Values such as those given in Table 4.5 are more difficult to obtain for plants growing in soil. Part of the problem arises because Equation 4.7 oversimplifies the system; dC/dz is not a simple function of $(C_r - C)$ because of the effect of v on C_r. For example, if all the ions which are absorbed are provided by a mass flow of the soil solution then $F = vC$, $C = C_r$, the diffusion term in Equation 4.7 is zero and it is acceptable. If, however, $vC < F$ then $C_r < C$ and the diffusion term becomes important in moving ions to the root. Conversely, if $vC > F$ the diffusion will be away from the root surface. In neither instance does C adequately describe the concentration of the ion in the solution arriving at the root surface (Fig. 4.3). Despite these objections, comparisons of predictions from Equation 4.7 and those obtained from a more complex, but more accurate, relationship have shown that the discrepancies are usually minor.

A slightly different, but essentially similar, equation has been used to analyse the uptake of potassium by leek seedlings growing in large pots of uniform soil (Table 4.6). The measured potassium concentration in the soil

Table 4.6. Nutrient and water flux data for roots of single leek plants during early growth. After Tinker, P. B. in Rorison (1969).

Age of plant (days)	Potassium uptake rate (nmol s^{-1})	Water uptake rate (mg s^{-1})	Root length (cm)	Surface area (cm^2)	Mean F (pmol cm^{-2} s^{-1})	Mean V (g cm^{-2} s^{-1})	VC/F (%)	Mean C_r (μmol ml^{-1})	$\bar{\alpha}$ (μm s^{-1})
49	0.8	19	630	150	5.6	13	10	0.24	230
62	3.1	31	2370	580	5.4	5.3	4	0.23	240
70	6.7	100	5340	1260	5.4	7.9	7	0.24	220
83	9.8	310	11750	1850	5.3	16	13	0.33	160

Mean potassium concentration C in bulk soil solution = 0.45 μmol ml^{-1}.

solution remained constant throughout the experiment and realistic values of D and ξ were used. The results were affected by some doubtful values for v at the first harvest but thereafter it appeared that the proportion of potassium supplied by flow to the root increased with time. The calculation of the mean concentration of potassium at the root surface, C_r, depends on the assumption that uptake was equally efficient over the whole root system, i.e. that $\overline{R\alpha}$ was essentially constant both over the whole root surface and over the period of the experiment. As was explained above, this is probably not really justified and if only part of the root system were absorbing potassium, C_r in that region would have been decreased and the contribution of diffusion to the flux in that region would have been increased. If either C_r or the proportion of the root system absorbing potassium had decreased then $\bar{\alpha}$ must necessarily have increased if the flux remained constant or increased. When C_r was zero then $\bar{\alpha}$ would have been infinite (Fig. 4.4). The observed fluxes were such that it is likely that more than about half the root system was absorbing potassium.

There is obviously a complex interaction between $\overline{R\alpha}$ in any one region of the root and C_r in the soil surrounding the root. When $\overline{R\alpha}$ is large it is possible that the movement of the ion through the soil may be too slow to maintain a constant C_r and the rate of uptake would therefore drop, and the contribution of diffusion to the movement through the soil would increase because of an increase in $(C-C_r)$.

Experiments somewhat analogous to but less comprehensive than that described above have been used to analyse the supply of other ions to plant roots. Accepting that there is a large effect of soil type, the environment and the plant studied, it has been found that copper and magnesium tend to be supplied mainly by the mass flow of the soil solution whereas potassium and phosphate are mainly supplied by diffusion. The depletion of phosphate from around the roots was shown qualitatively by some experiments in which wheat roots were allowed to grow through soil labelled with ^{32}P and

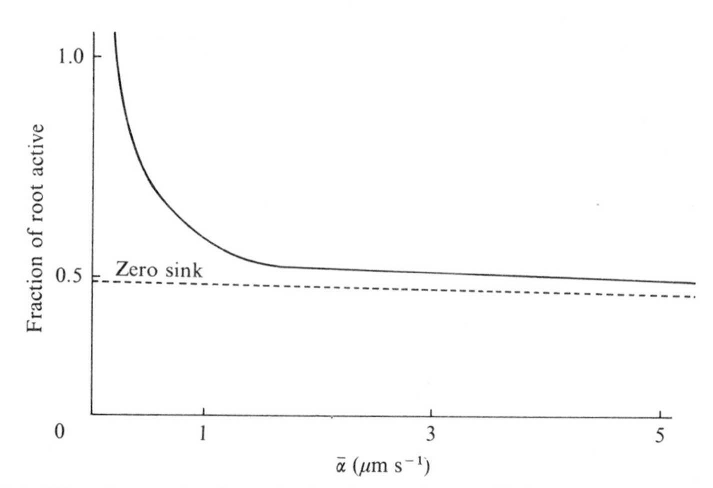

Fig. 4.4. The change in the calculated transfer coefficient of potassium into the root ($\bar{\alpha}$) with changes in the fraction of the total root length which absorbs potassium ions. The 'zero sink' line indicates that $\alpha = \infty$ and zero concentration of potassium at the surface of the absorbing roots. After Tinker, P. B. in Rorison (1969).

successive autoradiographs made of the root systems (Fig. 4.5). The development of a counting system with sufficiently precise resolution to allow quantitative measurements of this type would prove very useful. The supply of nitrate is complicated by the continual mineralization of organic nitrogen and reduction of the nitrate, but it seems likely that both diffusion and flow are important in the supply of nitrate.

The gradual depletion of ions from the whole mass of soil is a function of soil characteristics such as the buffering capacity (Fig. 4.6*a*), the relative rates of diffusion and mass flow of ions through the soil and many plant parameters, for example the rate of ion uptake and density of rooting (Fig. 4.6*b*) and the presence or absence of root hairs. Although root hairs significantly increase the surface area of the roots available for absorption, the restricted movement of ions through the soil usually results in the concentration of ions within the root-hair zone being depleted to a uniformly low level (cf. Fig. 4.7). The overall effect of the root hairs is therefore to increase the effective diameter of the root.

The further analysis of the absorption of ions from soil, and especially the effect of environmental factors such as the supply of water, requires a more detailed understanding of the processes governing ion uptake by roots. The total flux of ions into plant roots can be separated into two separate fluxes, one associated with the uptake of water by the plant, and the second independent of water uptake (Fig. 4.8). The ions passing through the root to the xylem can move almost entirely through the free space, i.e.

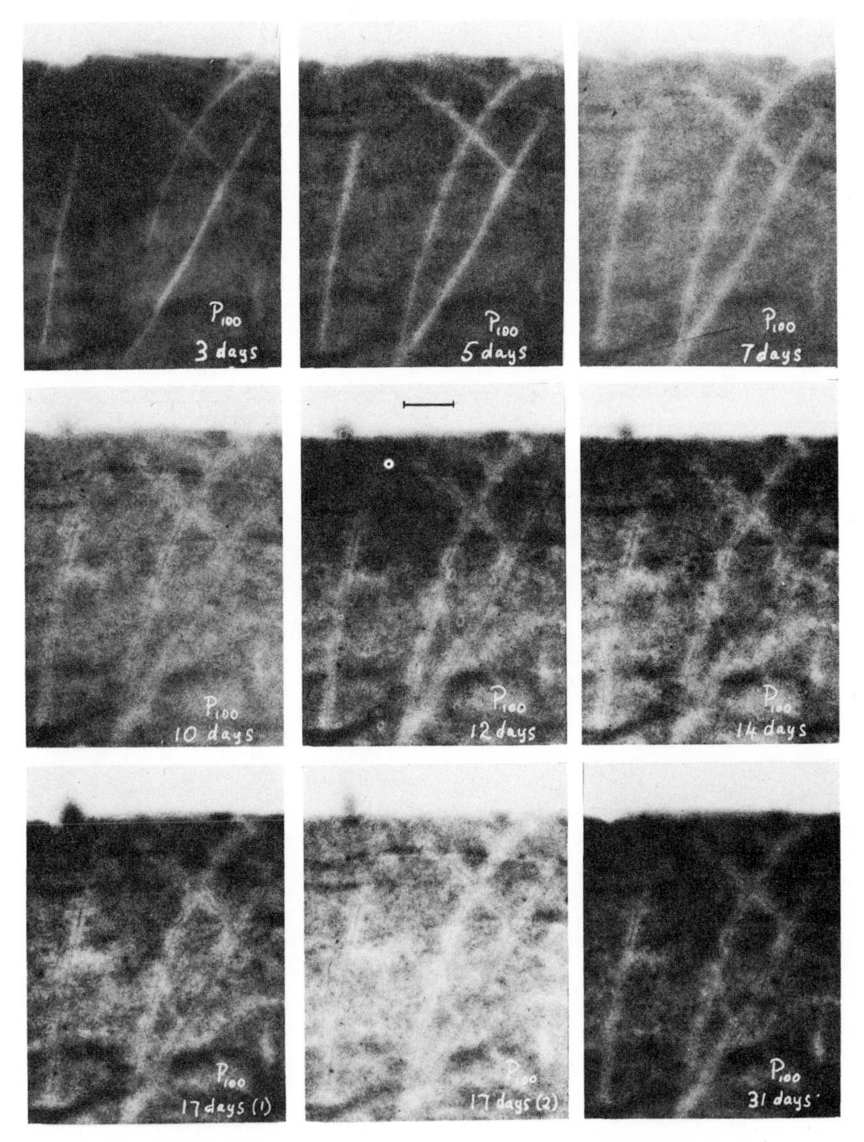

Fig. 4.5. Autoradiographs showing the depletion of ^{32}P from soil by wheat roots. After Lewis, D. G. & Quirk, J. P. (1967). *Pl. and Soil* **26**, 445–53.

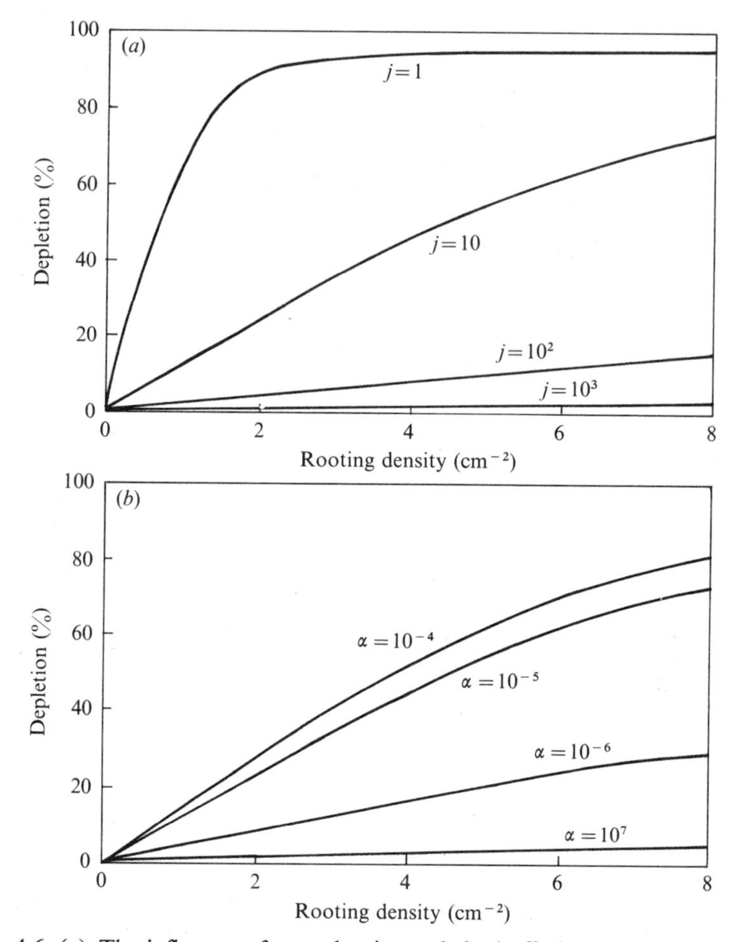

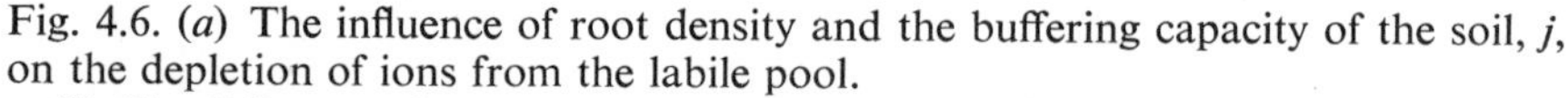

Fig. 4.6. (*a*) The influence of root density and the buffering capacity of the soil, j, on the depletion of ions from the labile pool.

(*b*) The influence of root density and ion transfer coefficient, α, into the root on the depletion of the labile pool. After Barley, K. P. (1970). *Adv. Agron.* **22**, 159–201.

those tissues external to the plasmalemmata. At some point, however, they have to move across a plasmalemma before they can enter the xylem. This transfer may occur in the outer tissues of the root, possibly at the endodermis, or even nearer to the xylem.

A kinetic analysis of the change in the rate of ion uptake at different concentrations of the external solution suggests the operation of two mechanisms with differing affinities for ions. In barley roots one of these – System I – has a low K_s for potassium of about 0.025 mM (K_s = the external

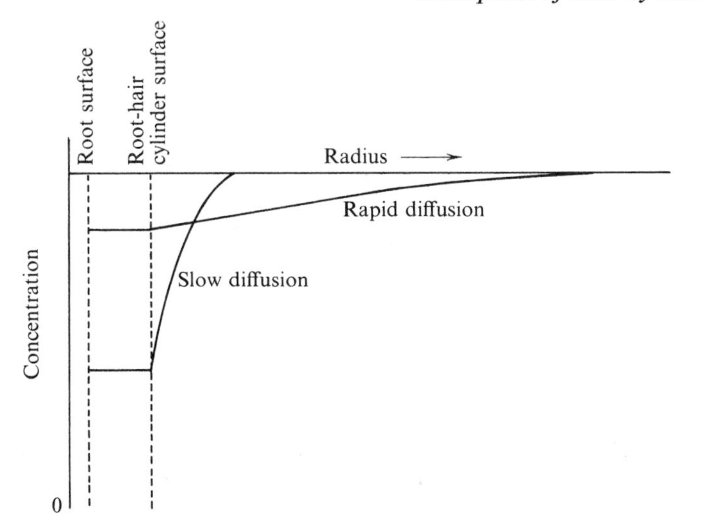

Fig. 4.7. Diagrammatic representation of the effect of root hairs on the concentration in the soil of ions with differing diffusion constants. After Nye, P. H. (1966). *Pl. and Soil*, **25**, 81–106.

concentration of ion when the rate of uptake is half maximum), that is, it has a high affinity for potassium. In *Ricinus* the K_s of the water-independent flux of potassium is very similar to that of System I and it has been suggested that the two are synonymous. System II has a K_s of about 17 mM, i.e. a low affinity for potassium although the maximum rate of uptake of both systems is about 12 mM g^{-1} fresh wt of root h^{-1}. System I is saturated at very low concentrations of potassium, less than 1 mM, which approximate to those found in the soil solution. It has been suggested that System I operates at the plasmalemma and System II at the tonoplast, although another suggestion is that both act in parallel at the plasmalemma. When the effect of the external concentration on the transport of ions to the shoot is analysed it appears to follow System I kinetics. This – together with the observation that ions which enter the vacuoles of the root cells take longer to move into the xylem than do ions which do not enter vacuoles – tends to support the suggestion that System II is concerned with transport into the vacuoles of root cells and that these are out of the main transport pathway. The half-time of turnover of 'cytoplasmic' phosphate is about 40 m compared with 90 h for 'vacuolar' phosphate.

The relative extent to which ions move through the free space or symplasm (interconnected cytoplasm) into the xylem, and eventually the aerial parts, appears to be a function of the rate of water uptake and the nutrient status of the plants. In plants of high nutrient status the total flux of ions appears to be dependent on the rate of transpiration, whereas in plants of low nutrient

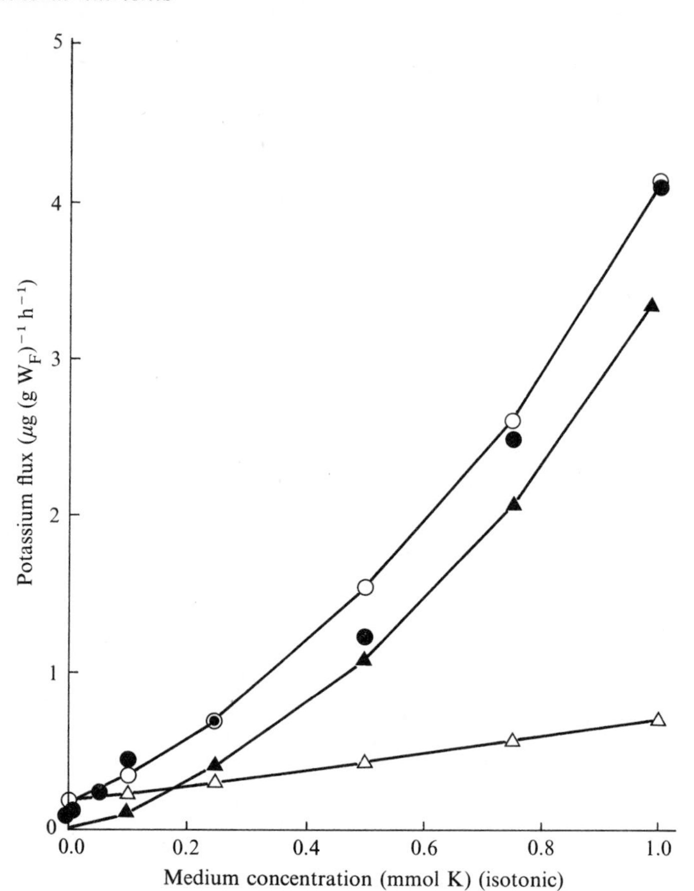

Fig. 4.8. The partitioning of the total flux (closed circles) of potassium ions into *Ricinus* roots into water-dependent (open triangles) and independent (closed triangles) fluxes. Open circles are the sum of the two fluxes calculated independently. After Baker, D. A. & Weatherley, P. E. (1969). *J. exp. Bot.* **20**, 485–96.

status the total flux is independent of transpiration. The reason for this difference appears to be that in plants of low nutrient status a large proportion of the ions are retained in the root tissues, i.e. the roots have first call on the ions that they absorb.

This passage through the free space, which contains many charged sites, and one or more charged membranes, allows ample opportunity for discrimination between ions. The degree of discrimination depends on the relative importance of the different pathways through the root; for example, the discrimination between Na^+ and K^+ is reduced when the rate of transpiration is increased.

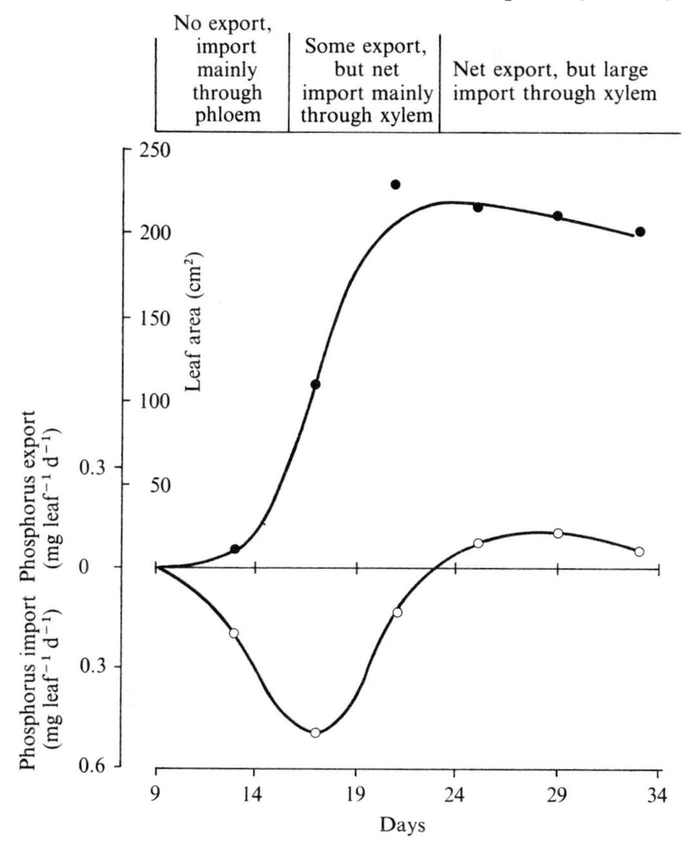

Fig. 4.9. The effect of leaf age on the increase in leaf area (closed circles) and changing patterns of phosphorus movement (open circles) into and out of the leaf. After Hopkinson, J. M. (1964). *J. exp. Bot.* **15**, 125–37; cf. also Greenway, H. & Gunn, A. (1966). *Planta* **71**, 43–67.

The delivery of ions to the aerial parts of the plant occurs initially in the xylem, with most of the ions moving in solution in the transpiration stream. The ions of the alkaline earth metals are different in that they appear to move along the walls of the xylem vessels by a series of ion exchange reactions.

Although many ions move through the xylem in the form in which they enter the plant, other ions are metabolized rapidly. For example, 30 per cent of the phosphate absorbed by barley plants is incorporated into nucleotides within 10 s and after 1 m 90 per cent is in some organic form. Similarly, there is evidence that the transported iron is complexed with organic acids. Nitrogen is usually absorbed as NH_4^+ or NO_3^-, but before it can be incorporated into amino acids, it must be reduced. The form in which it

Table 4.7. Percentage uptake of phosphorus from different compartments of the root zone by millet (*Pennisetum typhoides*) grown in Ghana at 122 cm × 122 cm spacing. After Nye, P. N. & Foster, W. N. M. (1961). *J. agric. Sci.* **56**, 299–306.

Age of crop (days)	Depth (cm)	Distance from base of plant (cm)			
		0–15	15–25	25–35	35–45
28	0–10	36	32	13	1
	10–20	4	6	1	2
	20–30	1	1	3	—
	30–40	—	—	—	—
	40–50	—	—	—	—
53	0–10	12	17	11	11
	10–20	11	10	8	6
	20–30	2	2	2	3
	30–40	1	1	1	1
	40–50	—	—	1	—
74	0–10	8	15	18	13
	10–20	7	7	7	5
	20–30	2	3	2	4
	30–40	1	1	2	2
	40–50	—	1	1	1
95	0–10	7	12	18	12
	10–20	4	7	5	6
	20–30	2	3	3	5
	30–40	1	2	3	3
	40–50	1	1	3	2

moves through the plant is dependent, therefore, on the site of nitrate reductase within the plant: in some species this is mainly in the roots, in others in the leaves. There is usually a correlation between the amount of ions delivered to an organ and its rate of transpiration. The organs which receive the most ions, the fully expanded leaves, have however a much smaller requirement than do young expanding leaves, the apices and developing fruit which depend to a large extent on the redistribution, in the phloem, of ions from the mature leaves (cf. Chapters 7 and 8). The change in the status of a leaf from a net importer of ions, mainly through the phloem, to an organ which imports ions through the xylem but is a net exporter through the phloem, is shown in Fig. 4.9. The ions of the alkaline earth metals are again different in that they seem to be unable to enter the sieve tubes, except with difficulty, and the concentration of these ions increases throughout the life of a leaf. Organs which grow rapidly, but with a restricted rate of transpiration, often suffer from Ca deficiency and show characteristic

Table 4.8. The percentage absorption of phosphate by ryegrass and a triploid hybrid between ryegrass and meadow fescue from 10, 30 and 60 cm under pure swards of ryegrass (R+R) and the triploid (T+T) and mixture of the two (R+T). The ^{32}P was injected into the soil on 20 May 1964 and 20 July 1964 and the 'hay' crop was harvested on 9 June 1964, 30 July 1964 and 10 September 1964, and the 'grazed' crop on 30 May 1964, 29 July 1964 and 20 August 1964; the latter was also harvested on 1 May 1964. Plants grown at Sutton Bonington, England. After O'Brien, T. A., Moorby, J. & Whittington, W. J. (1967). *J. appl. Ecol.* **4**, 513–20.

		Hay				Grazing			
Harvest	Depth (cm)	R(+R)	R(+T)	T(+T)	T(+R)	R(+R)	R(+T)	T(+T)	T(+R)
1	10	35.8	57.6	23.4	50.3	59.3	57.6	52.2	52.4
	30	60.5	41.3	70.1	48.6	39.9	40.9	47.1	45.7
	60	3.7	1.1	6.5	1.1	0.8	1.5	0.7	1.9
2	10	31.5	68.8	22.1	27.0	22.5	46.1	20.0	20.0
	30	50.6	25.6	32.1	17.4	51.2	43.8	26.2	20.0
	60	17.9	5.6	45.8	55.6	26.3	10.1	53.8	60.0
3	10	27.2	55.8	24.6	41.2	35.4	56.5	17.8	39.0
	30	26.8	26.0	27.5	9.8	39.6	25.9	35.6	26.1
	60	46.0	18.2	47.9	49.0	25.0	17.6	46.6	34.9

physiological disorders such as 'topple' in tulip peduncles, 'bitter pit' in apples, 'blossom end rot' in tomatoes, and 'tip-death' in etiolated potato and bean sprouts.

ABSORPTION OF MINERAL NUTRIENTS BY CROPS

The absorption of ions by plants growing in the field will be the resultant of all the factors discussed above. The roots of widely spaced plants extend both downwards and laterally in all directions away from the plant. The amount of soil available for exploitation will therefore be some function of the cube of the mean length of the main root axes. Since the younger parts of the root near the apices are the sites of most active absorption the regions of the soil from which absorption occurs will reflect the extension of the root system as well as the concentration in each region (Table 4.7). When inter-plant distances are small, for example in a closed grass sward, there is interference between the roots of adjacent plants and they do not spread as far laterally as with isolated plants. The presence of competing plants can affect, however, the depth from which plants absorb nutrients (Table 4.8).

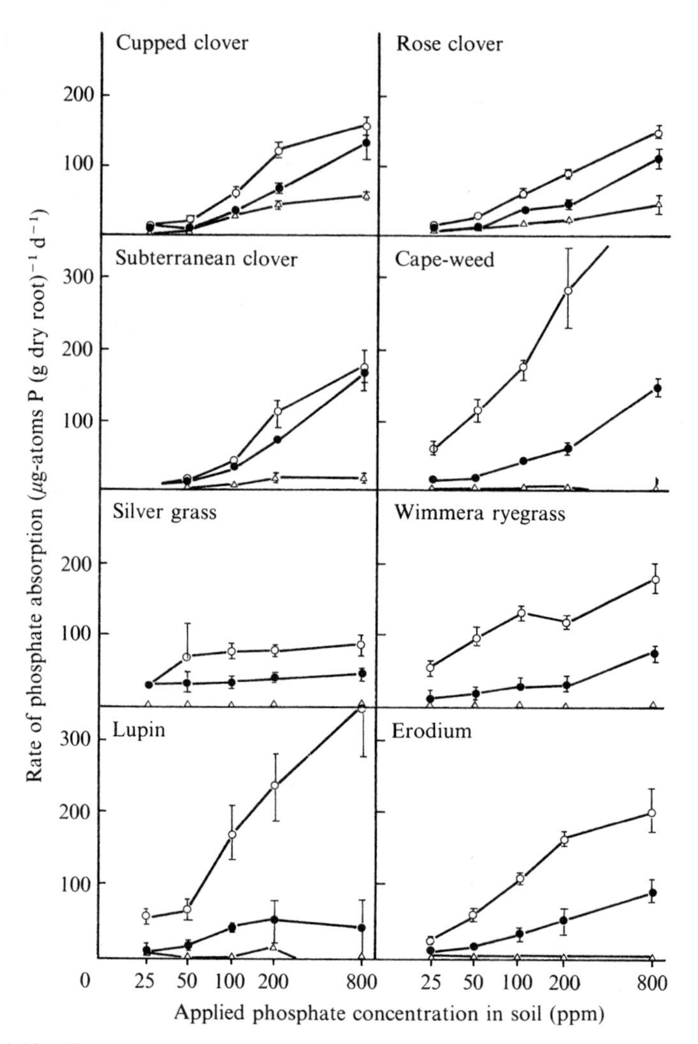

Fig. 4.10. The changes with time in the rate of absorption of phosphate by eight species from soil to which varying amounts of calcium dihydrogen phosphate had been applied. The vertical lines are ± SD of means of 4 replicates. After Keay, J., Biddiscombe, E. F. & Ozanne, P. G. (1970). *Aust. J. agric. Res.* **21**, 33–44.

The addition of fertilizer to a soil increases the concentration of ions in the labile pool and the rate of absorption of the ions (Fig. 4.10). As the fertilizer in the labile pool is removed the rate of absorption declines. In addition to these direct effects the complex chemical interactions in the soil often result in the addition of one fertilizer affecting the absorption of other ions (Fig. 4.11). The greater concentration of ions in the root resulting from

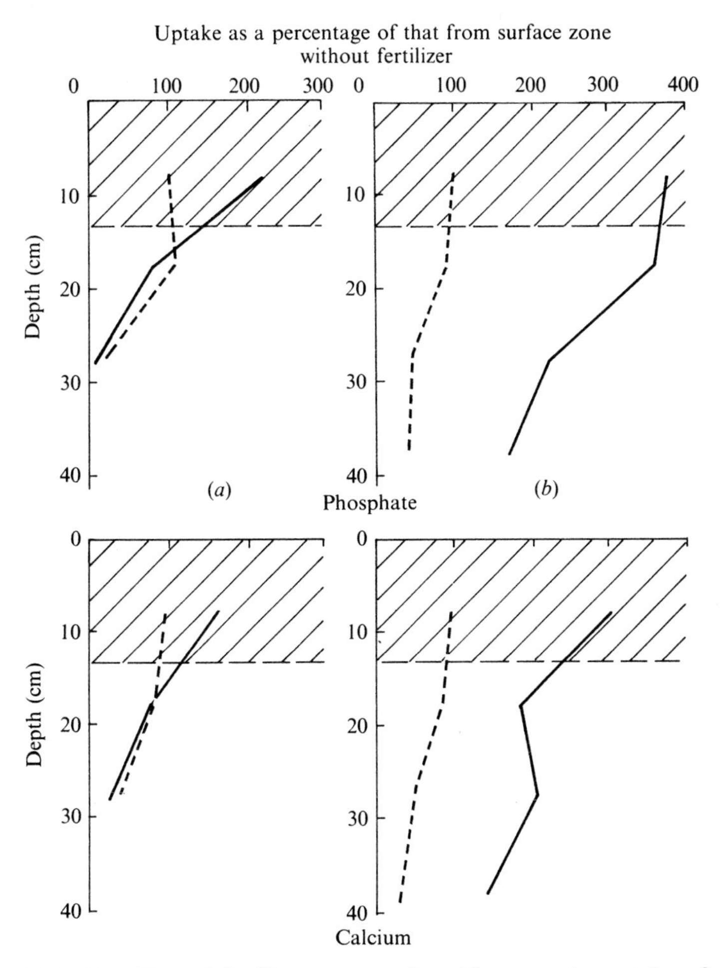

Fig. 4.11. The effect of fertilizer rotovated to 13 cm on the uptake of phosphate and calcium by kale from different depths in two soils: (*a*) a clay with flints; (*b*) a sandy soil. The results are expressed relative to the uptake from the surface zone in the absence of fertilizer. - - - - No fertilizer; —— fertilizer rotovated to depth indicated by shading (12.5 g N, 10.4 g P and 10.9 g K m^{-2}). After Russell, R. S. & Newbould, P. in Whittington (1969).

the increased uptake can lead to increased root growth and branching (cf. Chapter 7) and even more rapid uptake of ions. In some instances, however, the ionic concentration can be too high and the roots may then be damaged.

The influence of the soil water content on the absorption of nutrients is complex. There is an obvious effect on the mass flow of soil solution to the root surface and more indirect effects of the soil water content on the diffusion of ions. The drying-out of particular regions of the soil, such as the surface layers, may lead to compensatory growth in other regions, but the usual result is to reduce the uptake of nutrients from these surface layers. Since these regions are usually those with the greatest concentration of roots and nutrients the consequence is often severe, with the sparser deeper roots being unable to supply either sufficient water or nutrients to satisfy the needs of the crop. This can be seen even when there is an adequate supply of nutrients in the dry surface layers and the plants are obtaining sufficient water from the deeper regions.

FURTHER READING

Barley, K. P. (1970). The configuration of the root system in relation to nutrient uptake. *Adv. Agron.* **22**, 159–201.

Bartholomew, W. V. & Clark, F. E. (ed.) (1965). *Soil nitrogen.* Am. Soc. Agron., Madison, Wisc.

Milthorpe, F. L. & Moorby, J. (1969). Vascular transport and its significance in plant growth. *Ann. Rev. Pl. Physiol.* **20**, 117–38.

Nye, P. H. & Tinker, P. B. (1969). The concept of a root demand coefficient. *J. appl. Ecol.* **6**, 293–300.

Olson, S. R. & Kemper, W. D. (1968). Movement of nutrients to plant roots. *Adv. Agron.* **20**, 91–151.

Rorison, I. H. (ed.) (1969). *Ecological significance of the mineral nutrition of plants.* Blackwell Scientific Publications, Oxford.

Whittington, W. J. (ed.) (1969). *Root growth.* Proc. 15th Easter Sch. agric. Sci., Univ. Nottingham. Butterworths, London.

5

Photosynthesis and Respiration

PHOTOSYNTHESIS

Some 85–90 per cent of the dry matter of plants is carbonaceous material derived from photosynthesis, i.e. the light-dependent reduction of carbon dioxide. The overall process of photosynthesis can be considered in terms of three partial processes which are to varying extents interrelated and dependent on the internal and external environments.

The first of these processes is the diffusion of carbon dioxide to the chloroplasts, described by

$$P = -D\delta[CO_2]/\delta x = \Delta[CO_2]/r' \qquad 5.1$$

where $\Delta[CO_2]$ is the carbon dioxide concentration gradient and x the path length between the external air and the chloroplasts, and r' is the resistance to diffusion. The diffusion path can be divided into several sections; as will be seen below, and as observed for the diffusion of water vapour in the reverse direction (Equation 3.10), the most important resistance governing the diffusion is usually that associated with movement through the stomata. Environmental factors such as light intensity, and internal factors such as leaf water potential, influence stomatal aperture, and hence resistance, and must therefore be considered in this context.

The second process is photochemical, involving the interception of light by various pigments in chloroplasts. The light energy, with inorganic phosphate, P_i, is used in the splitting of water to produce molecular oxygen, reduced nicotinamide adenine dinucleotide phosphate, NADPH, and adenosine triphosphate, ATP, the overall reaction being

$$2H_2O+2ADP+4NADP+2P_i \xrightarrow[\text{chloroplast}]{\text{light}} O_2+2ATP+4NADPH \qquad 5.2$$

The process is controlled by the amount of radiation absorbed by the chloroplasts and is unaffected by the carbon dioxide concentration or the temperature.

In the third phase of photosynthesis the NADPH and ATP produced in the light are used to reduce carbon dioxide to carbohydrate and other compounds.

$$CO_2+2ATP+4NADPH \rightarrow (CH_2O)+H_2O+4NADP+2P_i+2ADP \qquad 5.3$$

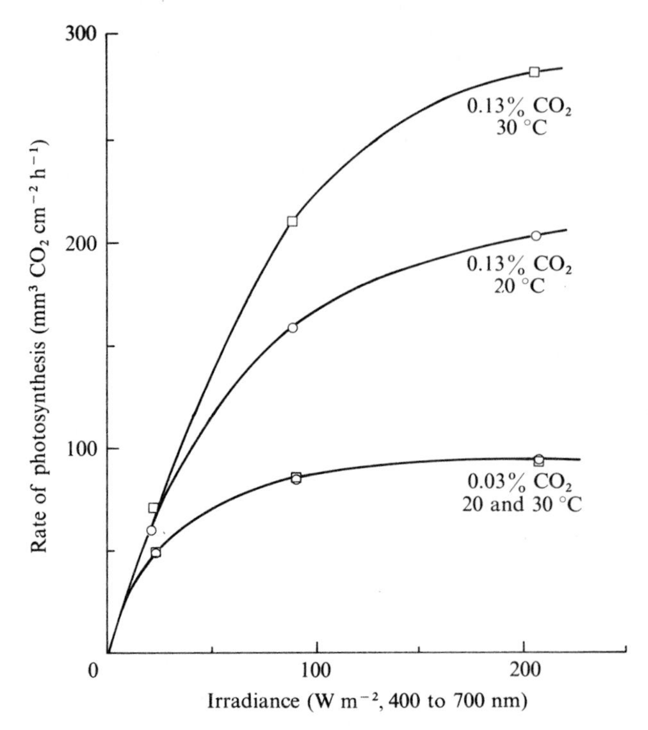

Fig. 5.1. The effects of irradiance, temperature and CO_2 concentration on the rate of photosynthesis. After Gaastra, P. (1962). *Neth. J. agric. Sci.* **10**, 311–24.

This reduction can proceed in the dark and is temperature dependent. The overall photosynthetic reaction can be written

$$CO_2 + 2H_2O \rightarrow (CH_2O) + H_2O + O_2 \tag{5.4}$$

The interaction between the three processes is illustrated in Fig. 5.1. At low irradiances the photochemical process is limiting and the rate of photosynthesis is dependent on the irradiance. As irradiance increases, the carbon dioxide supply becomes more important and eventually limiting. At the normal atmospheric concentration of carbon dioxide (0.03 per cent) there is little effect of temperature on the rate of photosynthesis. At high irradiances and a higher carbon dioxide concentration (0.13 per cent), temperature, and hence presumably the biochemical processes, become limiting and an increase from 20 to 30 °C produces a 50 per cent increase in the rate of photosynthesis. There are, of course, difficulties in interpreting overall responses such as these, especially as stomatal resistance increases with carbon dioxide concentration. However, with stomatal control eliminated and temperature optimal, similar limiting curves are found over a range of internal carbon dioxide concentrations from about 0.03 to 0.1 per cent, thus indicating that the carboxylation resistances have been saturated.

There are intermediate stages within this range of effects when two, or all three, of the partial processes can be operating at less than full efficiency. This situation is probably due to the complexity of the photosynthetic system in the leaf. Neither the length of the diffusion path from outside the leaf to the chloroplast, nor the irradiance at each chloroplast, is constant. It is possible therefore for a chloroplast situated near a stoma on the top of a leaf to have non-limiting supplies of carbon dioxide and radiation, whereas a chloroplast in the middle of the leaf may, at the same time, have insufficient radiation and carbon dioxide. Although this discussion has been limited to the photosynthesis of a single leaf, the same factors are important in consideration of the photosynthesis and growth of a crop.

Diffusion of carbon dioxide and photosynthesis

The diffusion of carbon dioxide from the atmosphere to the chloroplasts can be treated in a manner analogous to the diffusion of water vapour out of the leaf. Under steady state conditions, when the diffusion processes are limiting photosynthesis, the rate is given by

$$P = \{2(C_a - C_w)/(r'_a + r'_l)\} = \{(C_w - C_{chl})/r'_m\} = \{(C_{chl} - 0)/r'_x\} \quad 5.5$$

where P is the rate of photosynthesis, and C_a, C_w and C_{chl} are the concentrations of carbon dioxide in the atmosphere, at the mesophyll cell wall and at the chloroplast membranes, respectively. As with the transfer of water vapour (p. 37), here we express the resistances of the boundary layer and gaseous pathway in the leaf, r'_a and r'_l, per unit of leaf *surface* (i.e. twice the leaf area). The above expression holds when the resistances of the two sides are equal. It is much more difficult to identify the resistance to diffusion in the liquid pathway, r'_m, with leaf geometry, although more complete treatments are possible than are attempted here. It and the resistance reflecting the activity of the carboxylation reactions, r'_x, are here expressed per unit of leaf area. Both r'_a and r'_l are usually determined by measuring the comparable resistances to the loss of water vapour and taking $r'_a = 1.37r_a$ and $r'_l = 1.54r_l$.

The flux of carbon dioxide from the intercellular spaces in the leaf to the chloroplasts has two components, that from outside the leaf, given by the first term in Equation 5.5, and another due to carbon dioxide released in respiration. The second term in Equation 5.5 can, therefore, be written

$$F + nR = (C_w - C_{chl})/r'_m \quad 5.6$$

where F is the flux of carbon dioxide from outside the leaf, R is the respiratory flux of carbon dioxide, n that fraction of R which diffuses into the intercellular spaces, and $P = F + R$.

An alternative way of writing the third term in Equation 5.5 is

$$P = (bIC_{chl}/r'_x)/(bI + C_{chl}/r'_x) \qquad 5.7$$

where b is the maximum efficiency of conversion of radiation and I the incident radiation in the photosynthetically active range (400–700 nm).

The combination of these equations gives

$$F+R = \{C_a - F(r'_a + r'_s + r'_m) - nRr'_m\}/[1/bI\{C_a - F(r'_a + r'_s + r'_m)\} + r'_x] \qquad 5.8$$

the plot of which is a rectangular hyperbola with b the initial slope and the maximum value of $F = C_a - R(r'_x - nr'_m)/(r'_a + r'_s + r'_m + r'_x)$.

The net photosynthesis $F+R$ can therefore be considered in terms of the various resistances to carbon dioxide transport, the concentration of carbon dioxide in the air and the radiation. The effects of various internal and environmental factors on photosynthesis will now be considered in terms of these parameters.

Photochemical aspects of photosynthesis

The radiation incident on a leaf is either absorbed, reflected or transmitted through the leaf as discussed in Chapter 2. Only the absorbed radiation can be used in photosynthesis and the wave-length of this radiation is dependent on the various pigments in the plant, and is usually in the range 400–700 nm. The amount of radiation absorbed is a function of the concentrations of the pigments and their arrangement in the chloroplasts.

Although to this point light has been considered in terms of a wave motion, its absorption and utilization in photosynthesis are best considered in terms of the absorption of a series of particles or quanta. The energy content of a quantum is dependent on the frequency of the radiation, ν, (Table 5.1) and the relationship between them is given by

$$E = h\nu \qquad 5.9$$

where E is the energy content of the quantum (erg quantum^{-1}) and h is Planck's constant (6.625×10^{-27} erg s).

The efficiency of energy conversion in photosynthesis is best considered in terms of the quantum efficiency P/I_a where P is the rate of photosynthesis and I_a the number of absorbed quanta. Because the various pigments absorb radiation of different wave-lengths to differing extents, and because the energy content of the quanta in these wave-lengths varies, the quantum yield is a complex function of the wave-length of the radiation (Fig. 5.2).

There is evidence, outside the scope of the present discussion, that energy is transferred between pigments and the biophysics and biochemistry of the process are being actively investigated. For the present it is sufficient for us

Table 5.1. The properties of radiation of different wave-lengths.

Wave-length $(nm)^{-1}$	400	500	600	700
Wave number (cm^{-1})	25×10^3	20×10^3	16.7×10^3	14.3×10^3
Frequency (s^{-1})	7.5×10^{14}	6×10^{14}	5×10^{14}	4.5×10^{14}
Energy content per quantum (J $quantum^{-1}$)	5×10^{-19}	4×10^{-19}	3.3×10^{-19}	2.9×10^{-19}

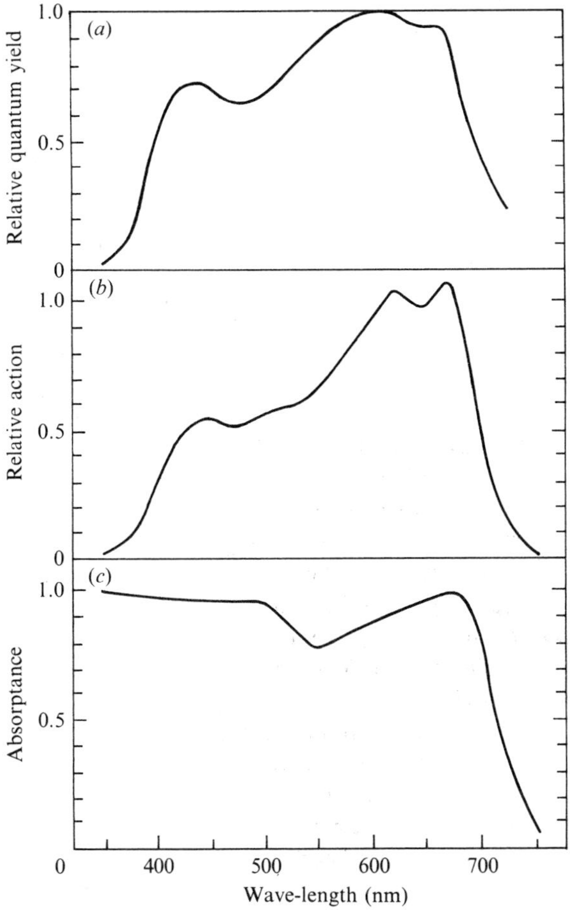

Fig. 5.2. Mean values obtained from 22 species grown in the field of (*a*) the mean relative quantum yield and (*b*) the mean relative action, both values normalized to a maximum value of 1.0 and (*c*) the mean absorptance of the radiation by the leaves. Action is defined as $a\ (C_L + C_D)/I$ where a is a constant to convert to micromoles per joule, C_L and C_D are the CO_2 fluxes into and out of the leaf in the light and dark respectively and I the irradiance. The quantum yield is given by b (action)/(wavelength $\times$ absorptance) where b is a constant to convert to moles per Einstein absorbed. After McCree, K. G. (1972). *Agric. Meteorol.* **9**, 191–216.

to accept that the light is used to produce ATP and NADPH according to Equation 5.2 and that these are then used in the reduction of carbon dioxide.

Reduction of carbon dioxide

The reduction of carbon dioxide appears to proceed by one of two main pathways, either of which can operate in the dark. For many years it was believed that the only pathway was one in which the carbon dioxide combined with the pentose sugar ribulose diphosphate to produce two molecules of phosphoglyceric acid and finally hexose (Fig. 5.3). This is the so-called Calvin cycle or C_3-pathway.

In the early to mid 1960s work in Hawaii showed that the first labelled compounds to accumulate after exposing sugar cane to ^{14}C-labelled carbon dioxide in the light were the 4-carbon-atom molecules oxaloacetate, malate and aspartate (Fig. 5.3). Subsequent work, mainly in Queensland, has shown that in this alternative fixation pathway, the Hatch–Slack or C_4-pathway, the carbon dioxide combines with phosphoenolpyruvate to produce oxaloacetate and the other C_4-compounds. The oxaloacetate can then take part in reactions leading to the formation of carbohydrates and other compounds and the regeneration of phosphoenolpyruvate. The exact pathway by which sugars are formed is not known. It appears likely, however, that carbon is transferred to phosphoglycerate by decarboxylation of oxaloacetate and refixation of the carbon dioxide liberated.

In addition, a third mechanism, known as Crassulacean acid metabolism, is common in a number of species including pineapple. Here, uptake of carbon dioxide occurs mainly in the dark, when stomata are open, resulting in the accumulation of organic acids, these being transformed to carbohydrates and other products during the day when the light reaction provides the necessary energy. There is little uptake of carbon dioxide during the day because of stomatal closure.

The C_4-pathway has been found to be quite common in species, especially grasses, of tropical and/or arid origins, whereas the C_3-pathway is usual in temperate species. Usually all members of a genus use the same fixation pathway, but this is not an absolute rule. *Atriplex hastata*, for example, uses the C_3-pathway whereas *A. spongiosa* and *A. nummularia* use the C_4-pathway.

A common feature of C_4-pathway plants is the presence of two different types of chloroplast. The mesophyll cells contain chloroplasts which resemble those in Calvin-cycle plants in size and in containing grana, but they do not accumulate starch. Arranged around the vascular bundles is a layer of cells, the bundle sheath cells, containing chloroplasts which are larger than those of the mesophyll cells and with few or no grana. Recent work on hybrids between *Atriplex* species with C_3- and C_4-pathways has shown that the F_1 hybrids appear to have characteristics intermediate between the

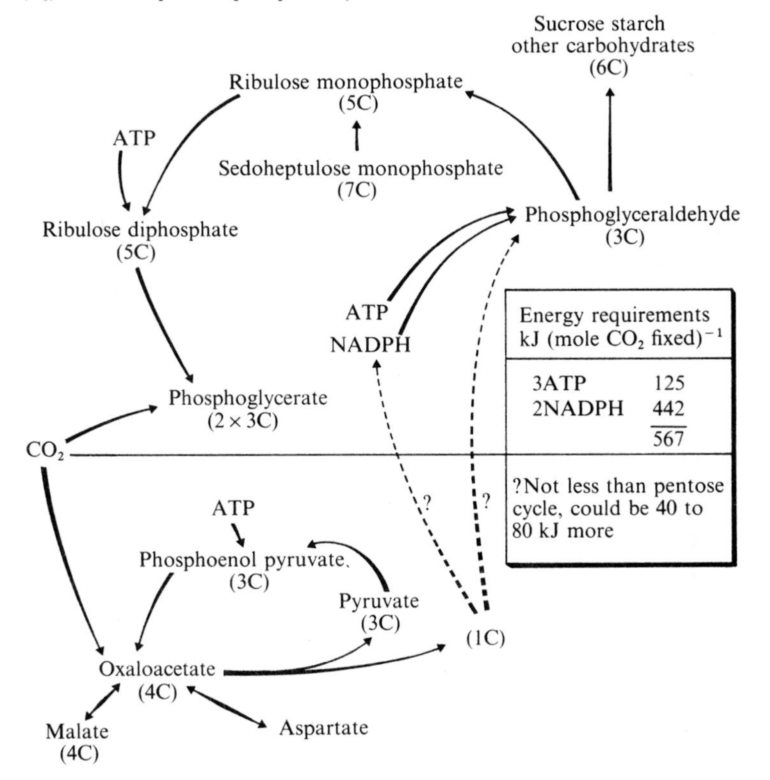

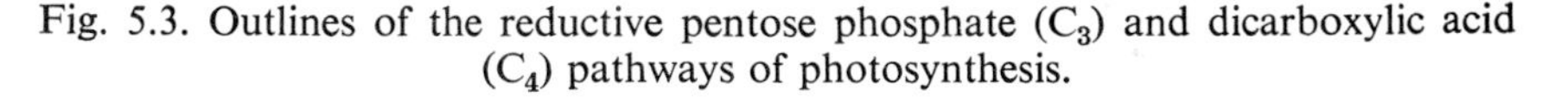

Fig. 5.3. Outlines of the reductive pentose phosphate (C_3) and dicarboxylic acid (C_4) pathways of photosynthesis.

parents with respect to their leaf structure and certain aspects of their physiology.

The operation of the C_4-pathway seems to require co-operation between both types of chloroplast. Only the bundle sheath chloroplasts contain enzymes capable of converting phosphoglycerate to hexose phosphate, the accumulated carbon being transferred from the mesophyll chloroplasts as malic or aspartic acid. This transfer probably takes place through the many plasmadesmata connecting the bundle sheath and mesophyll cells, but there are still a number of poorly understood issues associated with the quantitative aspects of this transfer.

In addition to the anatomical and biochemical differences between plants using the C_3 and C_4 carbon-fixation pathways, there are also physiological differences. The most important of these is that in C_3-pathway plants

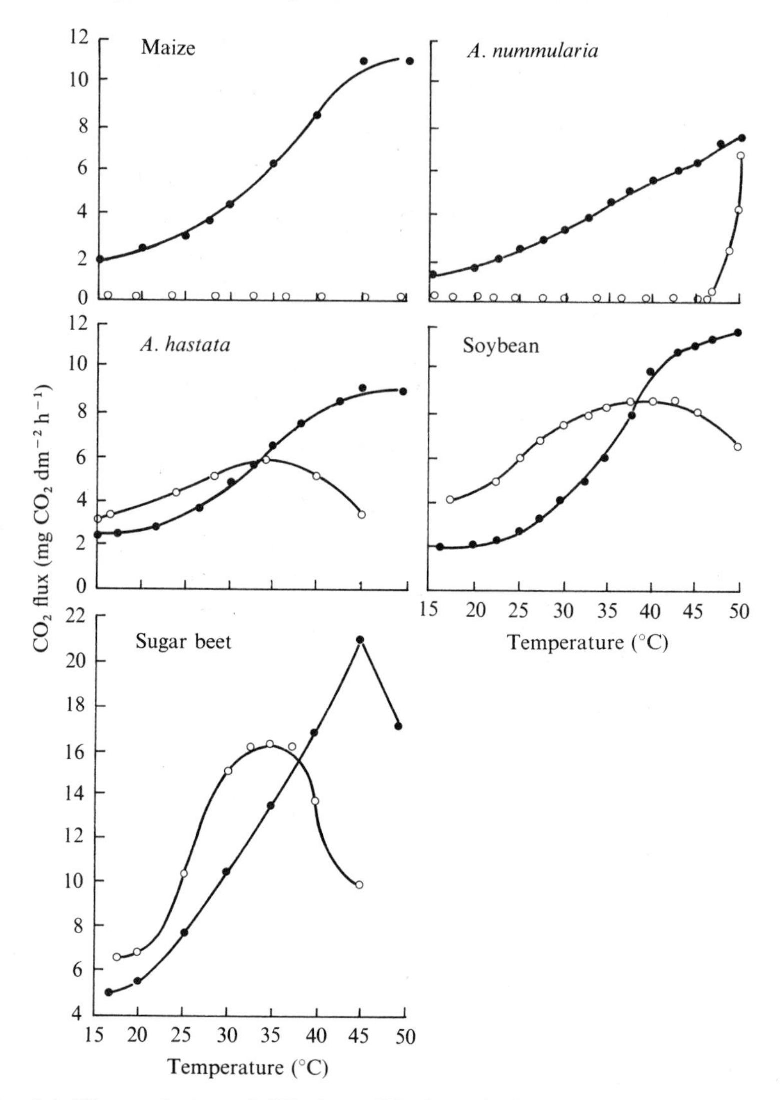

Fig. 5.4. The evolution of CO_2 into CO_2-free air from the leaves of five species exposed to an illuminance of 10000 fc (open circles) and in darkness (closed circles) at different temperatures. After Hofstra, G. & Hesketh, J. D. (1969). *Planta* **85**, 228–37.

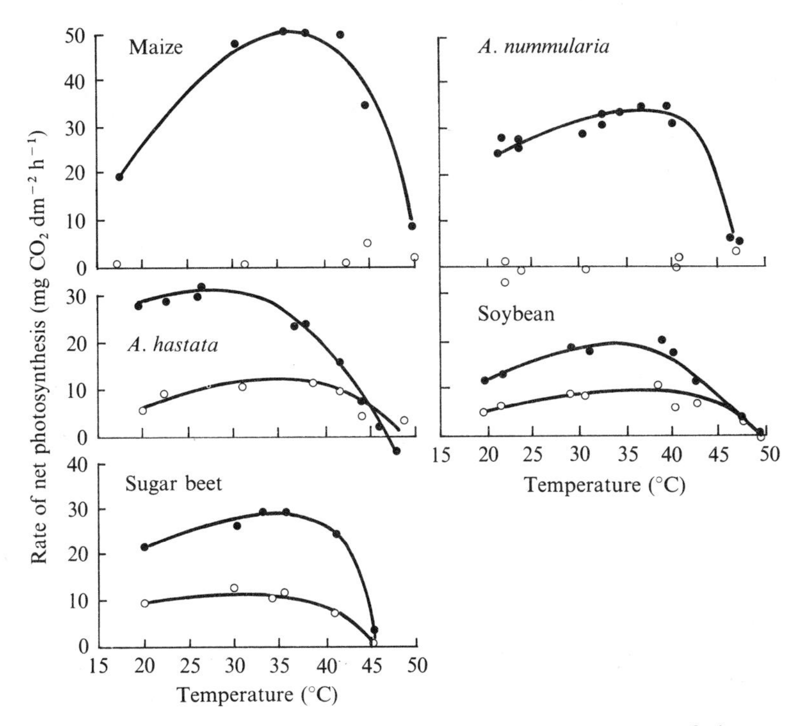

Fig. 5.5. The net photosynthesis (closed circles) and enhancement of photosynthesis in O_2-free air (open circles) of the leaves of five species at different temperatures. After Hofstra, G. & Hesketh, J. D. (1969). *Planta* **85**, 228–37.

between 20 and 50 per cent of the carbon fixed is immediately respired, the so-called photorespiration. In contrast, there is no detectable evolution of carbon dioxide in the light by C_4-pathway plants (Fig. 5.4). It is not known whether there is, in fact, no respiration in the light or whether the efficiency of carbon dioxide fixation in the mesophyll cells is so great that any carbon dioxide produced is immediately refixed. The absence of photorespiration in C_4-pathway plants results in higher rates of net photosynthesis (50–70 mg CO_2 dm^{-2} h^{-1}) than in C_3-pathway plants (15–35 mg CO_2 dm^{-2} h^{-1}). However, when photorespiration is inhibited, for example by a decrease in the concentration of oxygen, there is an increase in the rate of net photosynthesis of C_3-pathway plants (Fig. 5.5). This effect is not seen in C_4-pathway plants and so, at high light intensities and in the absence of oxygen, there may be no significant differences in the rates of net photosynthesis of plants using the two carbon fixation pathways (Fig. 5.6).

These results obtained in oxygen-free air show that photorespiration is aerobic, but it is not simply a reversal of the Calvin cycle nor does it use

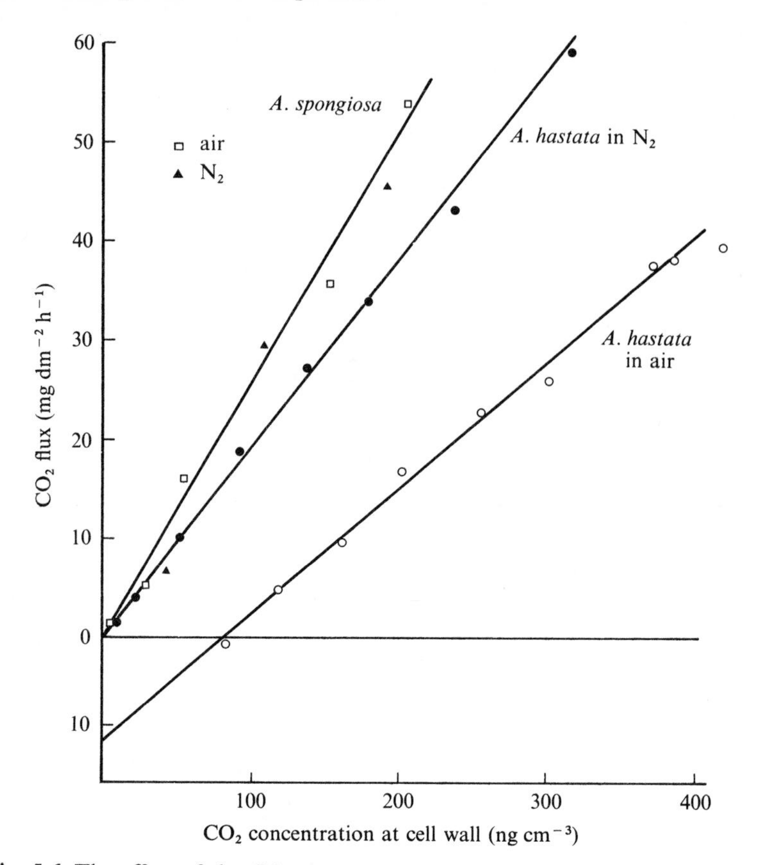

Fig. 5.6. The effect of the CO_2 concentration at the mesophyll cell walls on the rate of CO_2 exchange of well-watered *Atriplex spongiosa* and *A. hastata* at high irradiance and in normal and O_2-free air and constant leaf temperature. After Slatyer, R. O. in *The biology of Atriplex*, ed. Jones, R. (1969). CSIRO, Canberra, pp. 23–9.

the same substrates as dark respiration. The photorespiratory substrate appears to be glycollate which is formed in the light from intermediates of the Calvin cycle by both C_3- and C_4-plants. The synthesis and breakdown of glycollate occur in peroxisomes which are found throughout the mesophyll of C_3-plants, but are smaller and restricted to the bundle sheath cells of C_4-plants. If there is photorespiration in C_4-plants, therefore, there is a high probability that the carbon dioxide liberated will be re-assimilated since it would have to pass through the mesophyll, the site of initial carbon dioxide fixation, before it could leave the plant.

The operation of photorespiration has important effects on the energy requirements of photosynthesis and the resistances to carbon dioxide diffusion. The amount of energy contained by 1 mole of carbohydrate is about 490 kJ. However, the minimum amount required for the formation of 1 mole by the Calvin cycle is 567 kJ, and by the C_4-pathway some 40–80 kJ more (cf. Fig. 5.3). Although these energy requirements would appear to show that the C_3-pathway is more efficient, the efficiency is marred by the photorespiration which can increase the energy requirement for net photosynthesis by between 40 and 100 per cent.

These differences in photosynthetic efficiency have a significant effect on estimates of $r'_m + r'_x$, the minimum values for plants using the Calvin pathway being about 3–5 s cm^{-1} whereas those for plants using the C_4-pathway are about 1–2 s cm^{-1}. Which of these two resistances is the more important in determining the differences between C_3- and C_4-plants is still controversial. Some evidence suggests that the assimilation of CO_2 by phosphoenolpyruvate carboxylase is more efficient than by ribulose diphosphate carboxylase, and hence r'_x is lower in C_4-plants than in C_3-plants. Other methods of analysis, however, suggest that r'_m is the more sensitive resistance and that the main differences between C_3- and C_4-plants reside in the differing rates of transport of the carbon dioxide between the sub-stomatal cavities and the sites of fixation.

Although these effects have considerable fundamental interest, they play a minor part in determining the growth rates of whole plants. The effects discussed above are usually found in young leaves, and they become less obvious in older leaves. Further, when the growth of whole plants is considered, the proportion of dry matter used to produce new leaves and the ability of the plants to withstand, say, adverse temperature conditions or water deficits, tend to assume greater importance. There is often little difference between the maximum observed growth rates of C_3- and C_4-plants over short periods and even less when the comparison is made over longer periods or when single components of the total yield, such as grain, are considered (Table 5.2).

Environmental effects on photosynthesis

Light

The photochemical effects of light in photosynthesis have already been indicated, but the interpretation of the overall effect of light on the rate of photosynthesis is complicated by a variety of other factors. Light response curves such as those shown in Fig. 5.1 vary with the type of carbon fixation pathway used by the plant, C_3-plants becoming light-saturated at much lower light intensities than C_4-plants (Fig. 5.7). There are several possible reasons for this. The fact that there is light saturation implies that the limitation to photosynthesis is either in the supply of carbon dioxide to the chloroplasts or its reduction. The lack of photorespiration by C_4-plants is of obvious importance in this context, as are the usually smaller estimates of $r'_m + r'_x$.

Table 5.2. Comparison of carbon dioxide exchange properties of C_3- and C_4-species at seven levels of organization. The values quoted are amongst the highest ones in the literature (except for residual resistances which are the lowest); in general it is not difficult to find lower values for both species. An aspect not drawn out is the generalization that under cool, dull conditions C_3-species have a photosynthetic advantage over C_4-species and

	C_4-Species				
Level of organization	Species	Irradiance ($W\ m^{-2}$)	Photosynthetic parameter and its approximate value	Source	Comments
Primary carboxylation enzyme	*Zea mays*		$K_m(CO_2)$ $\approx 7\ \mu M$	19	Phosphoenolpyruvate carboxylase
			Residual resistance ($s\ cm^{-1}$)		
Leaf without stomatal effects	*Amaranthus viridis*	500	0.3	4	Approximate minimum values of
	Saccharum officinarum	500	0.3	4	photosynthetic potential with stomatal effects eliminated
			Net CO_2 exchange ($mg\ dm^{-2}\ h^{-1}$)		
Whole leaf	*Saccharum officinarum*	500	101	4	CO_2 exchange per unit leaf area at
	Pennisetum typhoides	600	101	17	near light saturation
Crop stand, gas analysis measurements	*Zea mays*	440	93.5	2	Transparent tent in field: L = 4.3, crop age 73 days
			Downward CO_2 flux ($mg\ dm^{-2}\ h^{-1}$)		
Crop stand, aerodynamic measurements	*Zea mays*	375	93.5–144	12	L = 4.2
	Zea mays	355	81	13	L = 4.5
		($cal\ cm^{-2}\ d^{-1}$)	Crop growth rate ($g\ m^{-2}\ d^{-1}$)		
Short-term crop growth rate	*Zea mays*	186	25	5	L = 4.0
	Zea mays	330	52 (incl. roots)	20	12-day mean L = 15 to 20
Long-term crop growth rate	*Pennisetum purpureum*		24	1	Total yield 87700 $kg\ ha^{-1}$ in 365 days
Grain growth	*Zea mays*		10.8 (of grain)	10	Total grain yield 14200 $kg\ ha^{-1}$ in 132 days

1. Anon. (1964). Dept Agric. Kenya Ann. Rep. 2.
2. Baker, D. N. & Musgrave, R. B. (1964). *Crop Sci.* **4**, 127–31.
3. Boerema, E. (1965). *Aust. J. expt. Agric. Anim. Husb.* **5**, 475.
4. Bull, T. A. (1969). *Crop Sci.* **9**, 726–9.
5. Buttery, B. R. (1970). *Crop Sci.* **10**, 9–13.
6. Cooper, T. G., Filmer, D., Wishnick, M. & Lane, M. D. (1969). *J. biol. Chem.* **244**, 1281–3.
7. Denmead, O. T. (1969). *Agric. Meteorol.* **6**, 357–71.
8. Dornhoff, G. M. & Shibles, R. M. (1970). *Crop Sci.* **10**, 42–5.
9. Evans, L. T. & Dunstone, R. L. (1970). *Aust. J. biol. Sci.* **23**, 725–41.
10. Hight, C. W. (1967). ASA Spec. Pub. No. 9, 87.

under hot, bright conditions the reverse is usual. The data in the table suggest that for plants under their own ideal conditions the advantage of C_4-metabolism is increasingly attenuated by other productivity-determining phenomena at higher and higher levels of spatial and temporal organization. Data compiled by R. M. Gifford.

	C_3-Species				
Species	Irradiance ($W m^{-2}$)	Photosynthetic parameter and its approximate value	Source	Comments	Ratio C_4/C_3
Spinacea oleracea		K_m CO_2 $\approx 450\ \mu M$	6	Ribulose diphosphate carboxylase	10–100
		Residual resistance ($s\ cm^{-1}$)			
Glycine max	335	1.6–3.5	8		5–10
Vigna luteola	380	2.5	15		
		Net CO_2 exchange ($mg\ dm^{-2}\ h^{-1}$)			
Typha latifolia	200	43–68	16		1.5–2.3
Triticum boeoticum	170	61–8	9		
Triticum aestivum	440	58	18	$L = 4.2$, crop age 70 days	1.6
		Downward CO_2 flux ($mg\ dm^{-2}\ h^{-1}$)			
Triticum aestivum	500	54	7	$L = 1.8$	1–1.7
Pinus radiata	250	126	7	$L = 4$	
	($cal\ cm^{-2}\ d^{-1}$)	Crop growth rate ($g\ m^{-2}\ d^{-1}$)			
Glycine max	186	15	5	$L = 4.0$	0.76–1.7
Helianthus annuus	150	68	11	$L = 4$–8	
Beta vulgaris		14	12	Total yield 42 500 $kg\ ha^{-1}$ in 300 days	
Oryza sativa		7.3 (of grain)	3	Total yield 13 800 $kg\ ha^{-1}$ in 190 days	1.5

11. Hiroi, T. & Monsi, M. (1966). *J. Faculty Sci., Univ. Tokyo*, Sect. III (Bot.) **9**, 242–85.
12. Inoue, E., Uchijina, Z., Udagawa, T., Horie, T. & Kobayashi, K. (1968). *J. Agr. Meteorol. Japan* **23**, 165–76.
13. Lemon, E. R. & Wright, J. L. (1969). *Agron. J.* **61**, 405–11.
14. Loomis, R. S. & Williams, W. A. (1963). *Crop Sci.* **3**, 67–72.
15. Ludlow, M. M. (1970). *Planta* **91**, 285–90.
16. McNaughton, S. J. & Fullem, L. W. (1970). *Pl. Physiol.* **45**, 703–7.
17. McPherson, H. G. (1970). Ph.D. Thesis, A.N.U., 1970.
18. Puckridge, D. W. & Ratkowsky, D. A. (1971). *Aust. J. agric. Res.* **22**, 11–20.
19. Waygood, E. R., Mache, R. & Tan, C. K. (1969). *Can. J. Bot.* **47**, 1455–8.
20. Williams, W. A., Loomis, R. S. & Lepley, C. R. (1965). *Crop Sci.* **5**, 211–15.

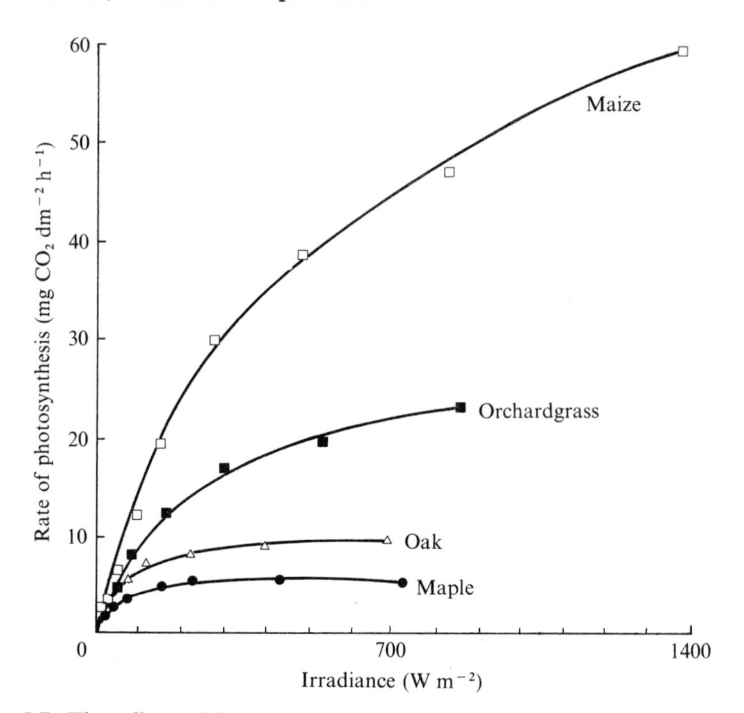

Fig. 5.7. The effect of irradiance on the rate of photosynthesis of four species. After Hesketh, J. D. & Baker, D. (1967). *Crop Sci.* **7**, 285–93.

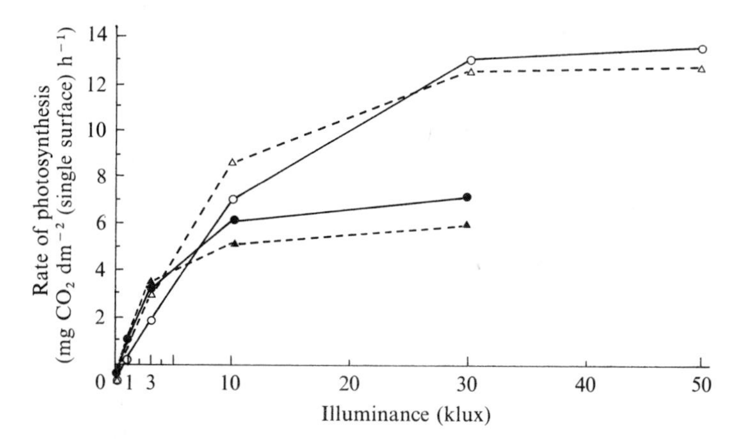

Fig. 5.8. The effect of irradiance on the rates of photosynthesis of sun (open symbols) and shade leaves (closed symbols) of *Quercus ilex* (circles) and *Fagus sylvaticus* (triangles). After Larcher, W. (1969). *Photosynthetica* **3**, 167–98.

A final possibility is that the stomata of C_3-plants are more sensitive to light than those of C_4-plants; although some evidence indicates trends in this direction, the variation between species within each group is probably greater than between groups. However, as was pointed out above, these differences do not appear to have a great effect on the rates of growth which can be maintained by the two types of plants, especially in the field situation.

Of greater interest in the latter situation are the different light response curves shown by sun and shade leaves (Fig. 5.8) or different races of the same species.

Carbon dioxide

The usual concentration of carbon dioxide in the atmosphere is 0.03 per cent and, as might be expected from Equation 5.1, an increase in the concentration to 0.1–0.15 per cent produces a two- to three-fold increase in the rate of photosynthesis (cf. Fig. 5.1). An example of the type of increases in yield obtained with *Callistephus chinensis* is given in Table 5.3. The increases in dry weight were due to an increase in net assimilation rate. Basically similar results have been obtained with field crops with some responding to rather higher concentrations of carbon dioxide; a ten-fold increase in carbon dioxide concentration having a similar effect to doubling the light intensity. It must be remembered, however, that it is not practicable to enrich the air with carbon dioxide except in enclosures such as a glasshouse; the much lower resistance to vertical transfer in the atmosphere than to diffusion into the leaf results in most of the added carbon dioxide being lost to the air above the crop.

The gradients in the carbon dioxide concentration could, theoretically, be influenced by changing the carbon dioxide concentration at the mesophyll cell wall or the chloroplasts. Neither of these two concentrations is amenable to manipulation. It is usually assumed that the concentration of carbon dioxide at the chloroplasts is zero. However, the few studies which have been made on the relationship between photosynthesis and the concentration of carbon dioxide at the cell wall (i.e. extensions of experiments such as that described in Fig. 5.6) show that the curves reach limiting values which lie in the range of 0.06 to 0.15 per cent. It would seem, therefore, that the transport system within the cell and/or the carboxylation enzyme systems are not adequate to handle concentrations much more than about twice that normally experienced. The carbon dioxide concentration at the mesophyll cell walls can be estimated by following the decrease in carbon dioxide concentration in a closed system around a photosynthesizing leaf. If light is not limiting, then at the compensation point the concentration of carbon dioxide outside the leaf should equal that at the assimilating cells. In Calvin-cycle plants the equilibrium concentration, Γ, reached in such a closed

Table 5.3. The percentage increases in dry weight of *Callistephus chinensis* at two times following enrichment of the atmosphere from 325 to 600 and 900 ppm. After Hughes, A. P. & Cockshull, K. E. (1968). *Ann. Bot.* **32**, 97–117.

Increase in CO_2 concentration	325 to 600	600 to 900	325 to 900
Expt 1 after 84 days' growth	32^{+++}	13^{+++}	73^{+++}
Expt 2 after 84 days' growth	28^{+++}	−1 n.s.	27^{+++}
Expt 2 after 115 days' growth	22^{+++}	4 n.s.	27^{+++}

(Significance levels: $^{+++}$ $P < 0.001$; n.s. not significant.)

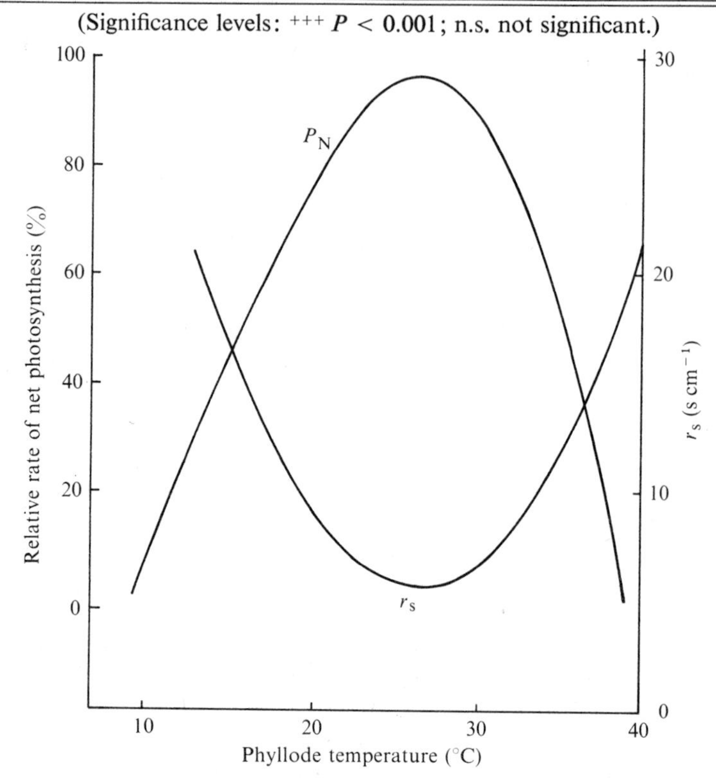

Fig. 5.9. The response of net photosynthesis and stomatal resistance to changes in the temperature of *Acacia harpophylla* phyllodes. After Van Den Driessche, R., Connor, D. J. & Tunstall, B. R. (1971). *Photosynthetica*, **5**, 210–17.

system is usually about 5–10 × 10^{-3} per cent, whereas with C_4-pathway plants it can reach 5–8 × 10^{-4} per cent. These are the minimum concentrations and are unlikely to be attained under normal growing conditions. One estimate for Γ in rapidly photosynthesizing beans is 18 × 10^{-3} per cent carbon dioxide but some arguments suggest that this is an overestimate and that in still air the usual minimum concentration may be approached.

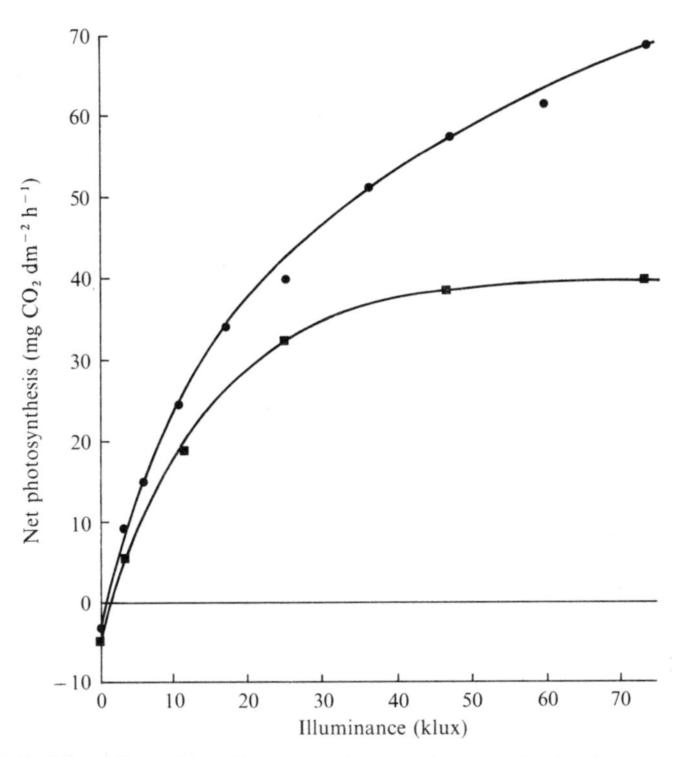

Fig. 5.10. The effect of irradiance on the net photosynthesis of *Panicum coloratum* grown at 20 °C and measured before (squares) and after (circles) overnight acclimatization at 30 °C. After Ludlow, M. M. & Wilson, G. L. (1971). *Aust. J. biol. Sci.* **24**, 1065–75.

The concentration of carbon dioxide in the intercellular space is the resultant of the depletion of carbon dioxide by photosynthesis and its evolution by photorespiration. Since in C_4-plants there is no evolution of carbon dioxide in the light the low values of Γ are not surprising. It follows from this argument that if the photorespiration of a C_3-plant were inhibited the value of Γ should decrease. This is confirmed by the results shown in Fig. 5.6, where the inhibition of photorespiration in *A. hastata* reduced Γ from about 0.008 ng cm^{-3} to zero.

Temperature

It is usually considered that the main effect of temperature on photosynthesis is exerted through effects on the biochemical reactions, and hence r'_m, but this is not always true. In *Acacia harpophylla*, for example, the main effect of temperature is on r'_s (Fig. 5.9) and there is a reasonable linear relationship

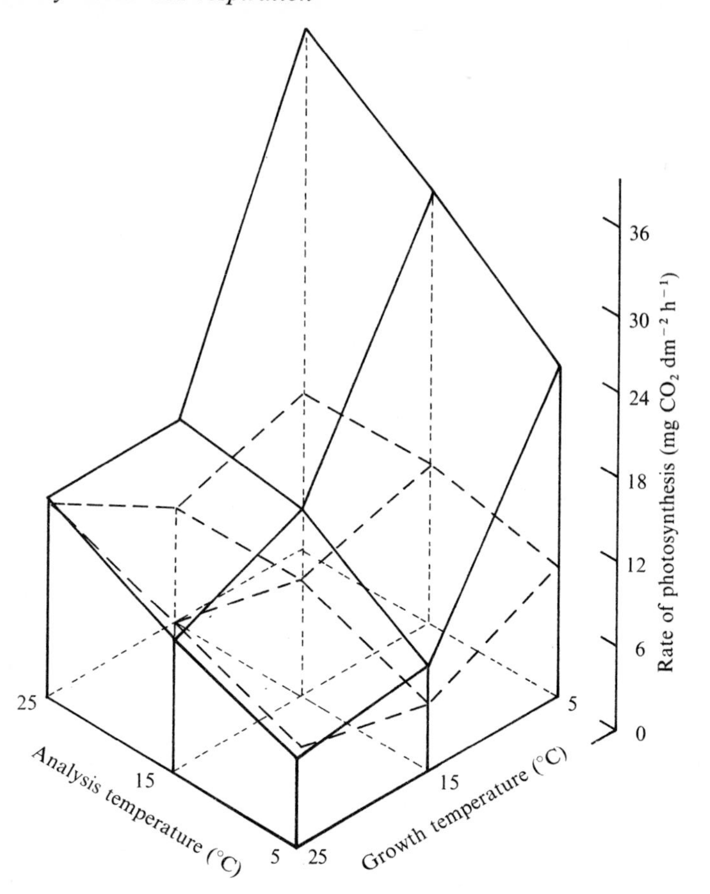

Fig. 5.11. The relationships between growth temperature, temperature during measurement and rate of photosynthesis of Norwegian (——) and Portuguese (- - - -) races of *Dactylis glomerata* plants grown at 100 W m^{-2} (300–700 nm) and measured at 240 W m^{-2}. After Treharne, K. J. & Eagles, C. F. (1970). *Photosynthetica* **4**, 107–17.

between the rate of photosynthesis and r'_s. In this species, changes in r'_s can account for 86 per cent of the variations in photosynthetic rate caused by changes in phyllode temperature.

The temperature optimum for *A. harpophylla*, 26 °C, is typical of most C_3-plants from both temperate and tropical regions, but the optimum of C_4-plants is usually about 35 °C. However, this optimum is usually not so pronounced and depends not only on the instantaneous effects of temperature but also on the temperature at which the plants are grown (Fig. 5.10). In addition, there can be an effect of the environment to which the plants have become adapted. For example, a race of *Dactylis glomerata* from

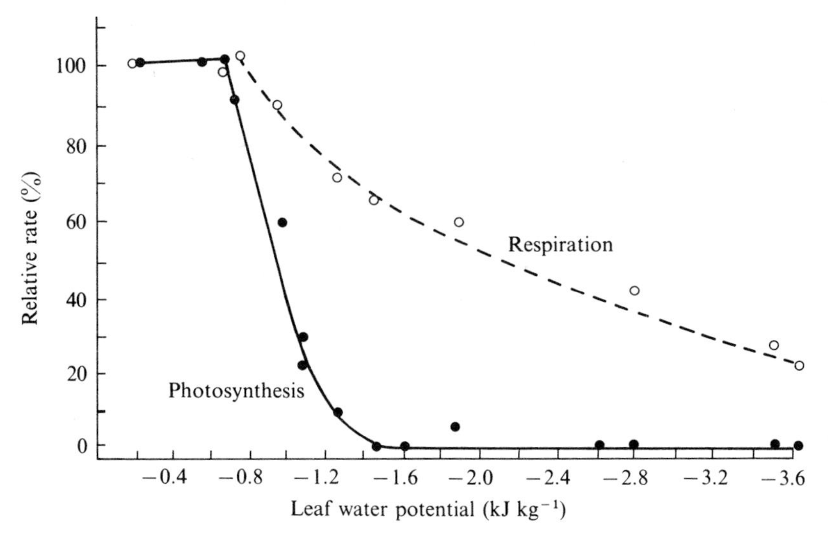

Fig. 5.12. The relationships with leaf water potential of the relative rates of photosynthesis (closed circles) and respiration (open circles). After Brix, H. (1962). *Physiologia Pl.* **15**, 10–20.

Norway had a much higher photosynthetic rate when grown at 5 °C than did a race of the same species from Portugal (Fig. 5.11) and, moreover, could respond more rapidly to an increase in temperature. Similarly, some tree species have a positive rate of net photosynthesis at −4 °C and can recover in a few days from exposure to much lower temperatures.

Water

Exposure of a plant to an increasing water deficit decreases the rate of photosynthesis relatively more than the rate of respiration, and there is a consequent decrease in the net rate of photosynthesis (Fig. 5.12). This effect has usually been interpreted as being mediated by an effect of the water deficit on r_s. For example, it is possible to induce changes in the relative water contents of leaves by reducing the root temperature. (This effect probably results from effects on root permeability which reduce the rate of water uptake.) A reduction of the relative leaf water content of cotton leaves below about 85 per cent caused an increase in the leaf resistance, whereas there was no effect on mesophyll resistance until the relative leaf water content was reduced to below 75 per cent, and even then r'_l was about three times greater than r'_m (Fig. 5.13*a*, *b*).

In this and other work it has been noticed that there can be regular oscillations in the rate of transpiration and photosynthesis which are a result of the rhythmic opening and closing of the stomata. Under these

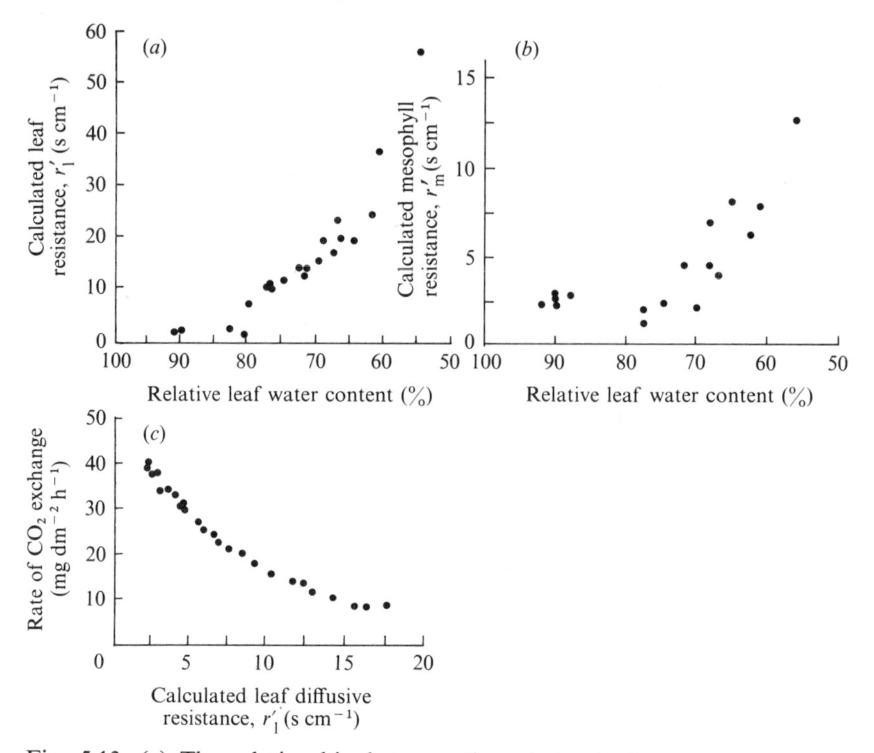

Fig. 5.13. (*a*) The relationship between the relative leaf water content (%) of cotton leaves and leaf resistance (r'_l) to CO_2 diffusion under steady state conditions. (*b*) The influence of relative leaf water content (%) on the mesophyll resistance to CO_2 transfer (r'_m) of cotton leaves. (*c*) The relationship between the leaf resistance (r'_l) and rate of photosynthesis of cotton leaves. After Troughton, J. H. (1969). *Aust. J. biol. Sci.* **22**, 289–302.

conditions the rate of photosynthesis is well correlated with the leaf resistance (Fig. 5.13*c*). This relationship has been shown to hold over a very wide range of phyllode water potentials in an Australian xerophyte *Acacia harpophylla* (Brigalow) (Fig. 5.14) but is not so clear-cut in some other species. In citrus, for example, there can be large changes in mesophyll resistance at quite high relative leaf water contents.

It is usually found that changes in stomatal resistance affect the rate of transpiration to a greater extent than the rate of photosynthesis. Theoretical analysis has shown that the relationship between the rates of transpiration and photosynthesis is a complex function of the temperature and rate of heat loss from the leaf and the boundary layer and mesophyll resistances. This analysis is beyond the scope of this work, but it does predict that there can be physiological advantages in variations in stomatal resistance on the same leaf and between different leaves; these have been observed.

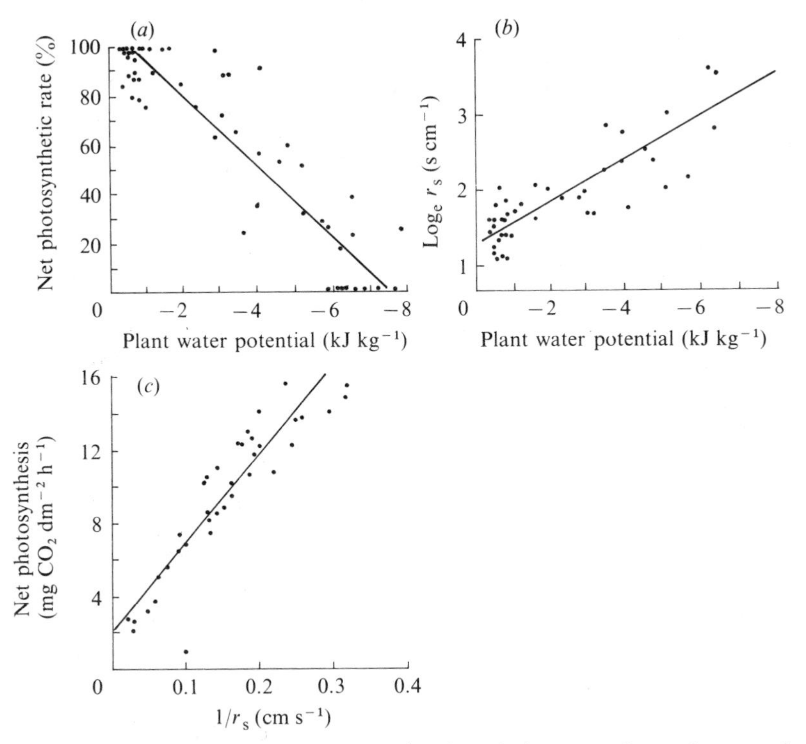

Fig. 5.14. The relationships between (*a*) the relative rate of net photosynthesis and plant water potential; (*b*) log of stomatal resistance to water vapour transfer and plant water potential; and (*c*) the rate of net photosynthesis and stomatal conductance for water vapour transfer ($1/r_s$) over the whole range of plant water potentials. After Van Den Driessche, R., Connor, D. J. & Tunstall, B. R. (1971). *Photosynthetica* **5**, 210–17.

A comparison between the rates of photosynthesis and transpiration in wheat and sorghum has been made at a series of light intensities and temperatures. Except at high light intensities, the stomatal resistance of sorghum tends to be higher than that of wheat and therefore the rate of transpiration tends to be lower. However, the lower Γ in sorghum tends to produce a greater rate of photosynthesis than in wheat and hence the rate of photosynthesis relative to that of transpiration is highest in sorghum (Fig. 5.15). This more efficient use of water may be one of the major advantages of plants having the C_4-photosynthetic pathway.

Mineral nutrients

There is usually a positive relationship between the supply of mineral nutrients and the rate of photosynthesis, which is exerted through an effect of the mineral nutrients on the mesophyll and carboxylation resistances

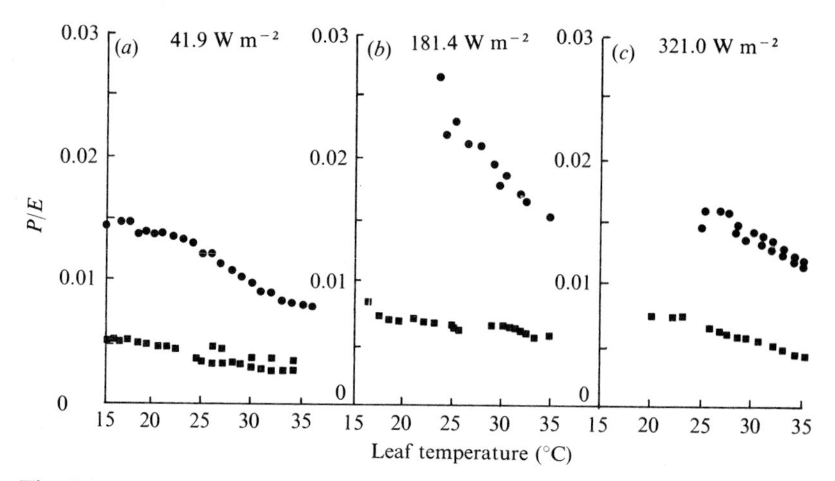

Fig. 5.15. The ratio of the rates of photosynthesis, P, and transpiration, E, of wheat (closed squares) and sorghum (closed circles) at a range of leaf temperatures and irradiances. After Downes, R. W. (1970). *Aust. J. biol. Sci.* **23**, 775–82.

(Fig. 5.16). Associated with this effect is a decline in the photosynthetic ability with increasing leaf age. The maximum rate of photosynthesis is usually found at about the time of maximum leaf expansion and this maximum rate is maintained for longer when nutrient supplies are ample. Eventually, however, the rate does decline, with the decline occurring sooner when nutrients are in short supply (Fig. 5.17) but eventually reaching the same value. The rate of respiration of a leaf also declines with age, but more slowly than true photosynthesis, and hence the decline in the rate of net photosynthesis is less than the former (Fig. 5.18). Since the leaf is growing throughout part of the period during which these changes are occurring, its requirements for assimilates also change and hence the leaf initially imports assimilates from older leaves and at about the time it reaches a third of its maximum area it starts to export assimilates whilst continuing to import them. It becomes a net exporter at about the time of attaining half the maximum area. As explained in Chapter 4, the pattern of import and export of some mineral nutrients follows a similar pattern, but with a different time scale and the onset of net export of phosphate from a leaf at maximum leaf area usually coincides approximately with the maximum rate of export of assimilates (cf. Fig. 5.18).

Effects of internal factors

In addition to the effects described above, which can be considered to result from changes originating in the leaves themselves, there is now much evidence which suggests that the rate of photosynthesis can also be influenced

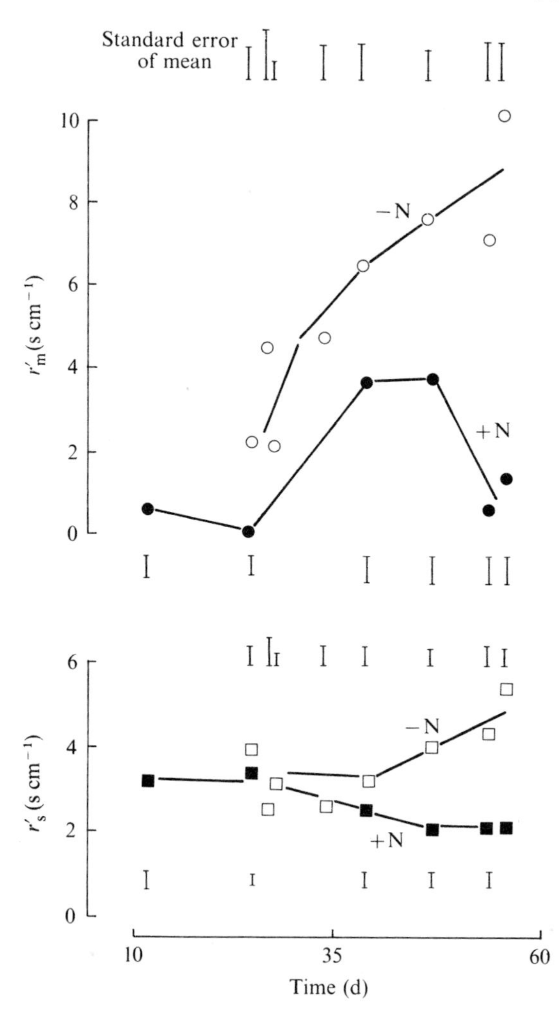

Fig. 5.16. The effect of increasing nitrogen deficiency (−N) of maize on the stomatal (r'_s) and mesophyll (r'_m) resistances to CO_2 transfer. After Ryle, G. J. A. & Hesketh, J. D. (1969). *Crop Sci.* **9**, 451–4.

by events occurring in parts of the plant remote from the photosynthetic regions. This phenomenon is probably best documented in potatoes. In this species, removal of tubers or inhibition of their growth by lowering the soil temperature leads to a reduction in the net assimilation rate or rate of photosynthesis. The reverse treatment, i.e. a reduction in the size of the photosynthetic system by partial defoliation, has no effect on the rate of growth of the tubers, presumably because of an increase in the photosynthetic efficiency of the remaining leaves. Similar effects have also been found after

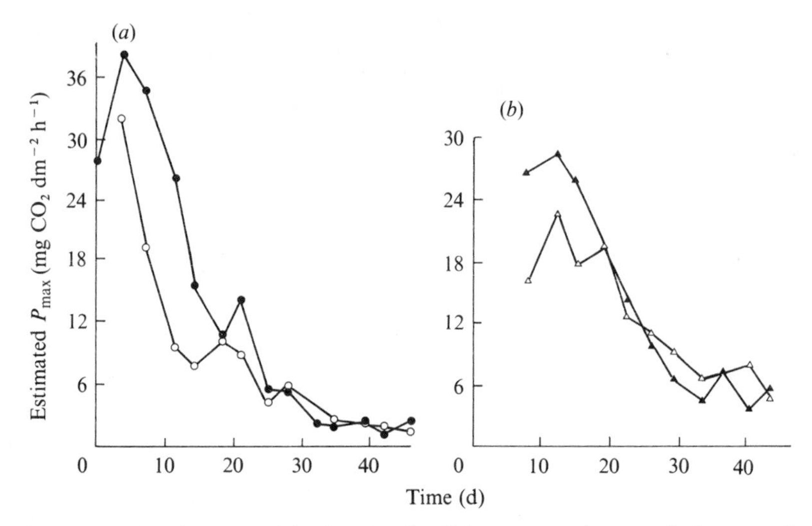

Fig. 5.17. The changes with time in the light-saturated rate of photosynthesis of (*a*) leaf 3 and (*b*) leaf 6 of tomato plants grown in 12.5-cm (open symbols) and 17.8-cm (closed symbols) pots arranged to supply restricted and adequate supplies of mineral nutrients. After Peat, W. E. (1970). *Ann. Bot.* **34**, 319–28.

the partial defoliation of apples, tomatoes and lucerne. In the lucerne experiments the increase in the rate of photosynthesis of the stubble leaves following defoliation was associated with a decrease in $r'_m + r_x$ (Fig. 5.19).

The importance of potato tubers in controlling photosynthesis has also been inferred from ontogenetic changes in both the rate of photosynthesis and the net assimilation rate of potatoes. The net assimilation rate increases at the time of tuber initiation, then falls until about the time of maximum leaf area and finally rises again. Analogous increases in the net assimilation rate have been found at the onset of fruit growth in peppers and at the end of the growing season in tomatoes, peas, soybeans and *Callistephas chinensis*. This later increase is difficult to explain, coming as it does at a time when leaf area and incoming radiation are normally decreasing and when the increasing age of the leaves remaining on the plant would normally be expected to lead to a decrease in the rate of photosynthesis. The most reasonable explanation would seem to be that the continued rapid growth of the tubers or other 'sinks' causes an increase in the rate of photosynthesis of the remaining leaves.

A different approach to this problem has been the use of reciprocal grafts between partners with contrasting properties. For example, the net assimilation rates of reciprocal grafts between sugar beet, which has a large storage root, and spinach beet, with no storage root, were greatest when the sugar beet was used as the stock; the source of the scion leaf had little effect.

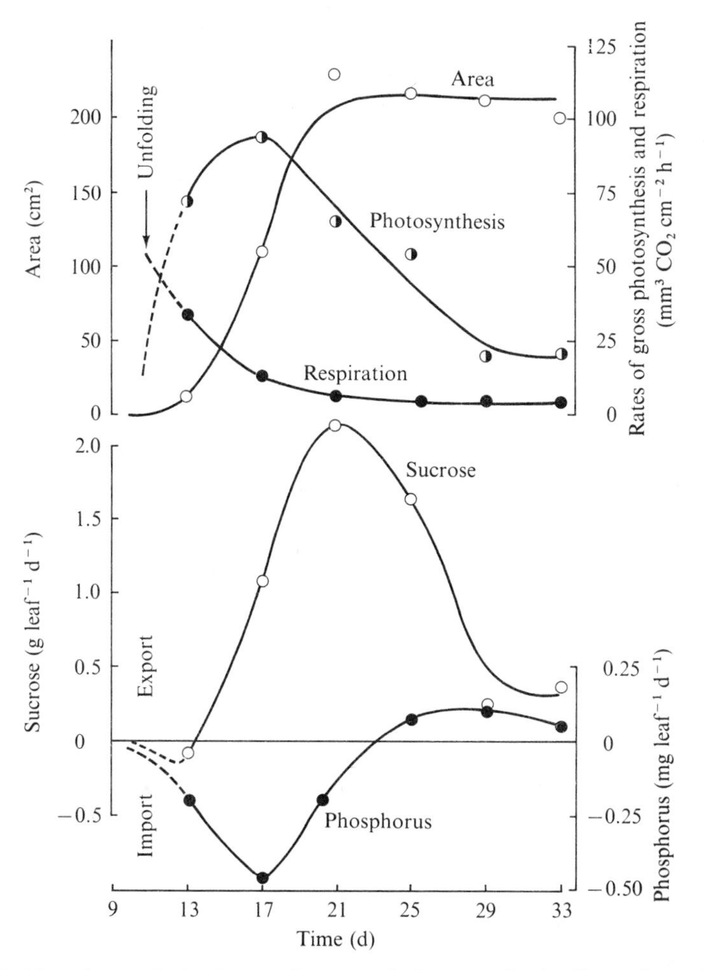

Fig. 5.18. The change in leaf area, L, rates of photosynthesis, P_s, and respiration, R, and import and export of sucrose, S, and phosphorus, P, during the ontogeny of the second leaf of cucumber. After Hopkinson, J. M. (1964). *J. exp. Bot.* **15**, 125–37.

Similarly, in grafts between two varieties of potatoes the yield of tubers and their starch content were independent of the type of scion material used. This type of experiment was carried even further by using reciprocal grafts between tomato and potato. In these plants a greater proportion of assimilated ^{14}C went into the fruit and tubers when tomato scions were grafted on to potato stocks.

An interaction between source and sink has also been found in wheat, where removal of the ear leads to a 40 per cent reduction in the net

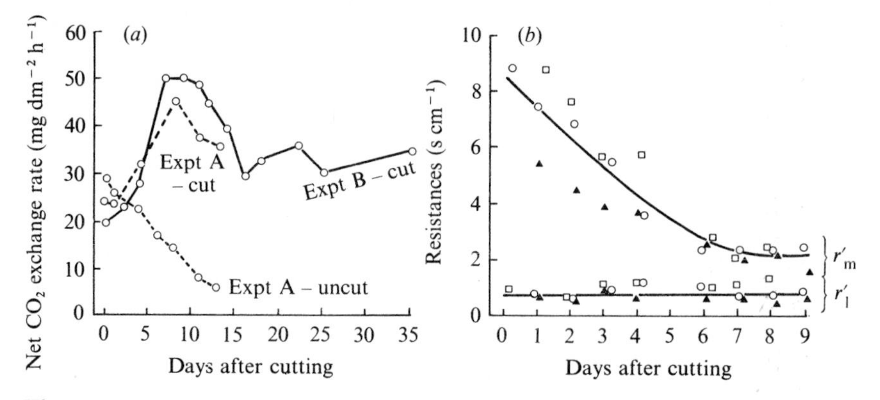

Fig. 5.19. Changes in (*a*) the net rate of CO_2 exchange of stubble leaves and (*b*) the leaf, r'_l, and mesophyll, r'_m, resistances to CO_2 diffusion following partial shoot removal. The results of two experiments are shown in (*a*) and a third, but comparable, experiment in (*b*). After Hodgkinson, K. C., Smith, N. G. & Miles, G. E. (1972). *Aust. J. agric. Res.* **23**, 225–38.

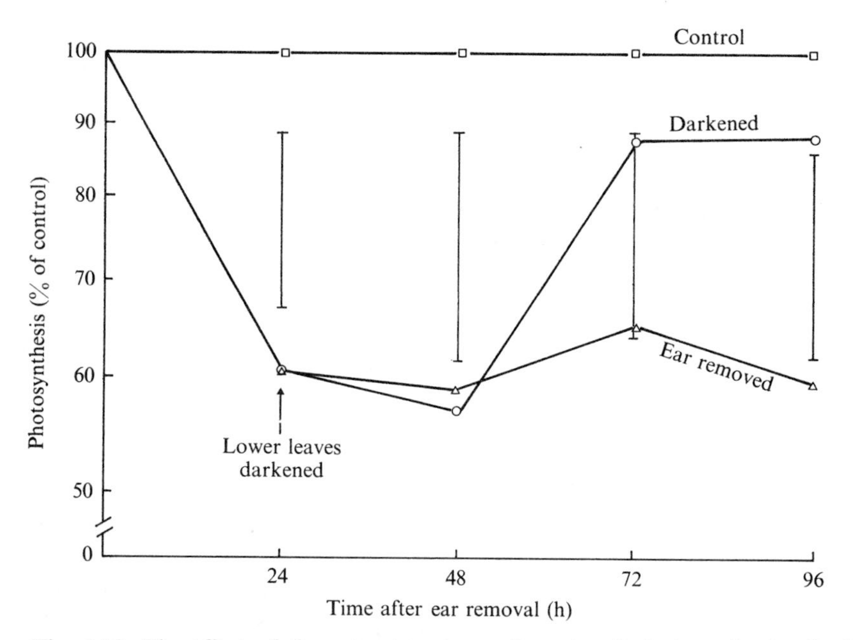

Fig. 5.20. The effect of three treatments on the rate of photosynthesis of wheat plants: the ear removed at zero time; the ear removed at zero time and the lower leaves darkened 24 h later and control plants. Each point is the mean of 4 replicates and the vertical lines indicate the 5 % least significant differences for each time. After King, R. W., Wardlaw, I. F. & Evans, L. T. (1967). *Planta* **77**, 261–76.

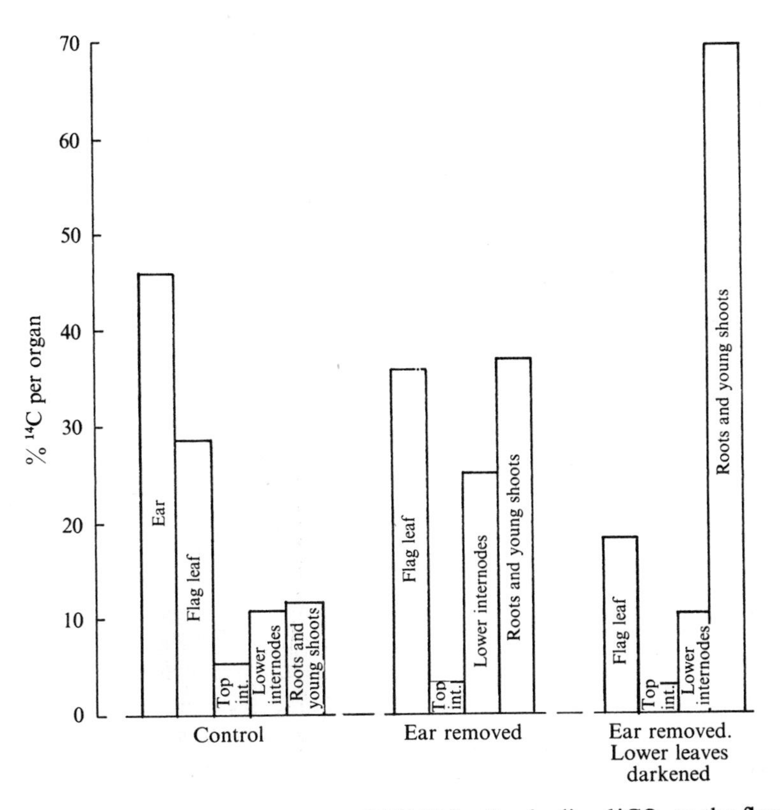

Fig. 5.21. The distribution pattern of ^{14}C 24 h after feeding $^{14}CO_2$ to the flag leaves of wheat plants: control plants, ear removed 24 h before exposure to $^{14}CO_2$ and lower leaves darkened 24 h after ear removal and 3 days before exposure to $^{14}CO_2$. After King, R. W., Wardlaw, I. F. & Evans, L. T. (1967). *Planta* 77, 261–76.

photosynthesis of the flag leaf. This effect was partially reversed when the other leaves on the plant were darkened (Fig. 5.20).

This apparent feedback of information from sink to source appears to occur where the plant cannot develop sufficient alternative sinks to utilize any surplus assimilates. For example, potatoes cannot produce sufficient lateral shoots to compensate for the tubers that are removed, whereas in *Chrysanthemum morifolium* removal of the major sinks, the flower heads, merely leads to a diversion of material into the roots and no detectable drop in the net assimilation rate. Similarly, if the initials of Jerusalem artichoke tubers are removed, the assimilates are diverted into the roots. The situation in wheat appears to be midway between these two extremes. As was seen above, removal of the main sink, the ear, caused a reduction in the rate of photosynthesis of the flag leaf. At the same time there was an increase in

Table 5.4. The effect of continuous light on the photosynthesis of sugar cane leaf discs floating on sugar solutions in Warburg flasks and supplied with $^{14}CO_2$ from ^{14}C-labelled carbonate–bicarbonate buffer and carbonic anhydrase. After Waldron, J. C., Glasziou, K. T. & Bull, T.A. (1967). *Aust. J. biol. Sci.* **20**, 1043–52.

Treatment	Initial rate of gross photosynthesis ($\mu l\ O_2\ h^{-1}$)	Rate of photosynthesis after exposure to light for 22 h ($\mu l\ O_2\ h^{-1}$)	Percentage inhibition of gross photosynthesis
Control	52	39	25
0.4 M sucrose	78	59	24
0.4 M glucose	69	54	22
0.4 M mannitol	55	41	25

the proportion of the assimilates being translocated to the roots and young tillers. When the photosynthesis of the usual suppliers of assimilates to these organs, the older leaves, was inhibited, there was an increased export from the flag leaf (Fig. 5.21).

The way in which sink activity can control the rate of photosynthesis is not known. The complexity of the metabolic pathways involved in carbon fixation, and the multiplicity of the products of fixation, make the accumulation of any one product to inhibitory levels extremely unlikely. For example, it was not possible to inhibit the photosynthesis of sections of sugar cane leaf by floating them on sucrose solutions for 22 hours (Table 5.4) and a control mechanism involving end-product inhibition is, therefore, improbable. It is possible, however, that there are rapid changes in the activity of some carboxylating enzymes in response to treatments similar to those described above. For example, the activity of ribulose-1,5-diphosphate carboxylase increased following both the partial defoliation of *Phaseolus* and *Zea* and the growth of plants at high light intensities. However, in some situations, the effect is primarily due to stomata.

Photosynthesis of plant communities

Diurnal trends

There are several techniques which can be used to measure the rate of photosynthesis of a leaf or plant in a controlled environment room. The measurement and analysis of the photosynthesis of plants and communities in the field are, however, considerably more difficult because all the environmental and internal factors discussed above operate simultaneously. Probably the most immediately obvious effect in this context is that the lower leaves

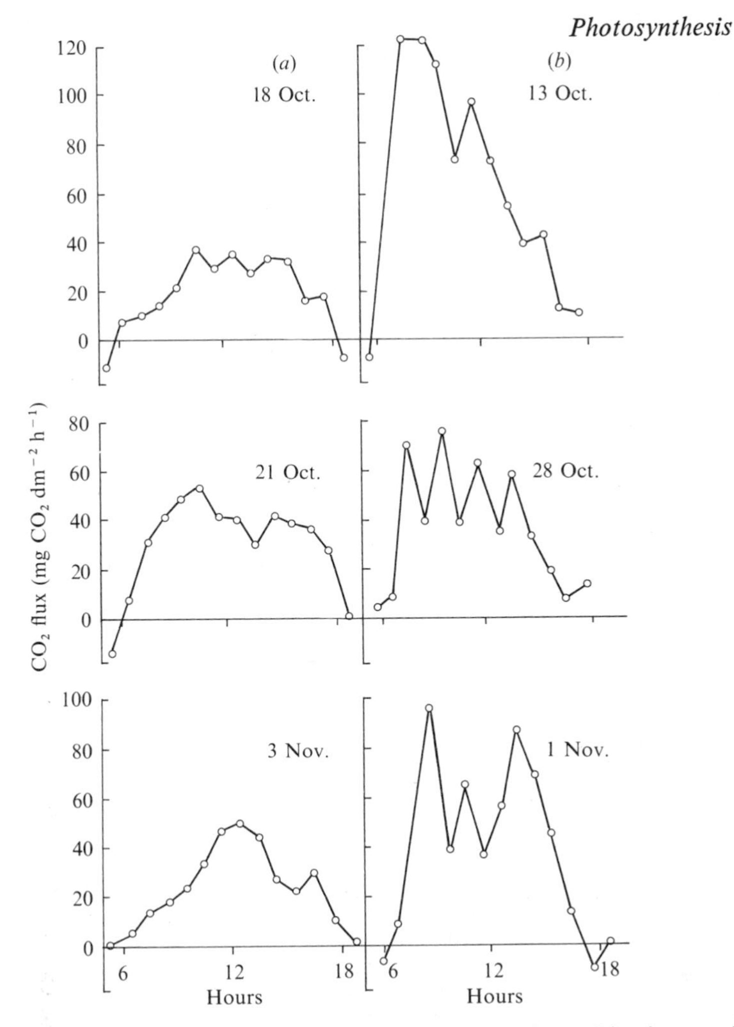

Fig. 5.22. The hourly mean downward fluxes of CO_2 above (*a*) wheat and (*b*) a pine forest. Both sites near Canberra, Australia. Negative values indicate a net upward flux. After Denmead, O. T. (1969). *Agr. Meteorol.* **6**, 357–71.

in a canopy are shaded and therefore have a lower rate of photosynthesis than the leaves at the top of the canopy. This effect is accentuated by the increasing age of the lower leaves and the export of mineral nutrients, which results in a higher $r'_m + r'_x$ and is associated with the onset of senescence. There are also effects caused by gradients of temperature and carbon dioxide concentration through the canopy.

An estimate of the average rate of photosynthesis of a crop can be obtained by enclosing part of the crop in a transparent tent and following the change in carbon dioxide concentration of the air inside the tent. This technique

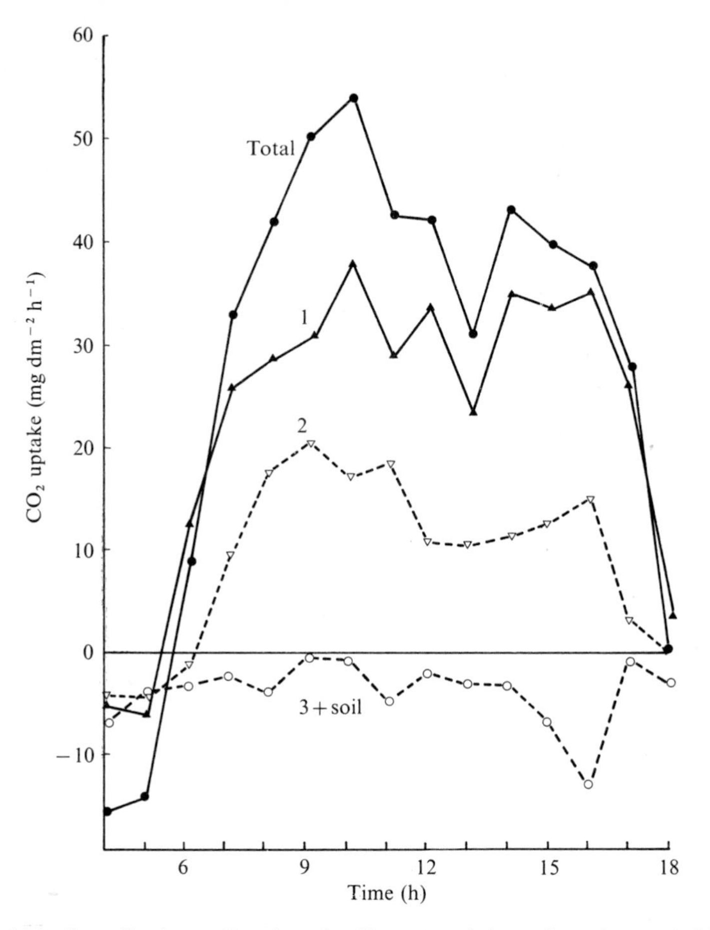

Fig. 5.23. Contributions of various leaf layers and the soil to the total CO_2 uptake of a wheat crop grown near Canberra, Australia. The measurements were made on 21 October (cf. Fig. 5.22). The total leaf area index of the crop was 1.67 and of layers 1, 2 and 3: 0.57, 0.58 and 0.52 respectively. Layer 1 was the uppermost layer. After Denmead, O. T. (1966). *Proc. WMO Seminar on Agricultural Meteorology*, pp. 445–82.

requires sophisticated air-conditioning equipment because the tent has to be ventilated and the carbon dioxide concentration maintained near ambient. It suffers from the disadvantage that the transfer processes between the atmosphere and crop cannot be the same as outside the enclosure. An alternative method is to use micrometeorological techniques to obtain the change in carbon dioxide concentration above and within the crop canopy. It is then possible to calculate the flux of carbon dioxide by applying Equations 2.9 to 2.12 described on pp. 28–9. The transfer coefficient for

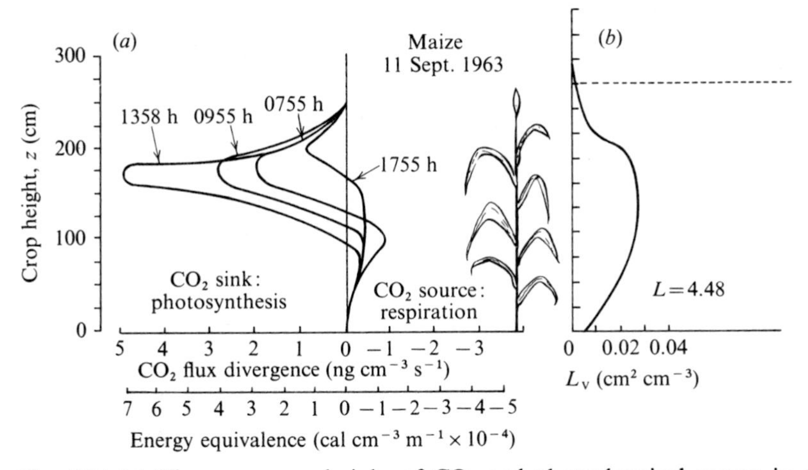

Fig. 5.24 (*a*) The sources and sinks of CO_2 and photochemical energy in a corn crop at Ellis Hollow, New York; negative and positive values indicate net photosynthesis and respiration respectively. (*b*) The vertical distribution of leaf area density in the same crop. After Lemon, E. R. & Wright, J. C. (1969). *Agron. J.* **61**, 405–13.

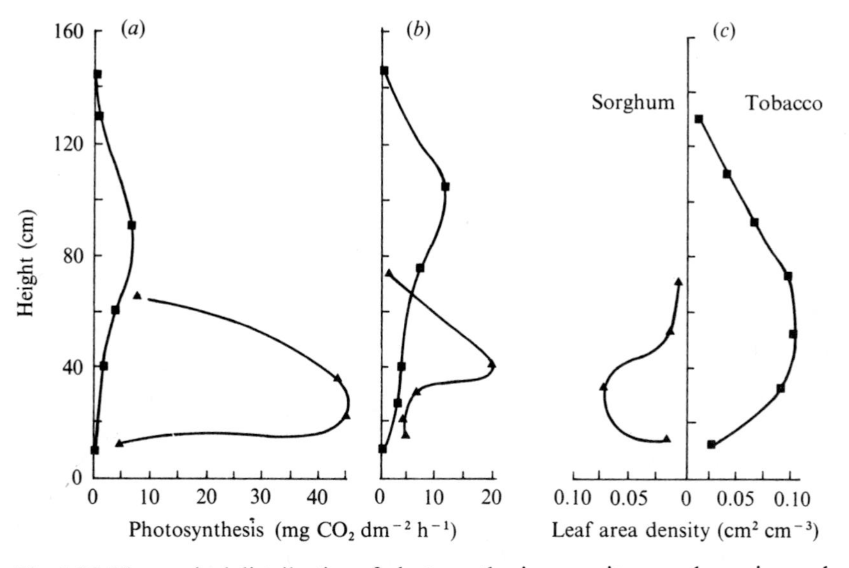

Fig. 5.25. The vertical distribution of photosynthesis per unit ground area in sorghum (triangles) and tobacco (squares) at (*a*) noon and (*b*) 4 h from noon, and (*c*) the vertical distribution of leaf area density of these crops grown at New Haven, Conn. After Turner, N. C. & Incoll, L. C. (1971). *J. appl. Ecol.* **8**, 581–91.

carbon dioxide, K_P, is assumed to be equal to either K_W or K_M and one of these is determined from the measured gradients of temperature and humidity or wind speed, respectively. This method was used to obtain the results shown in Fig. 5.22. The data for the pine forest are more variable than those for wheat, probably because the crop surface was rougher and the site was more sloping and had a shorter fetch relative to plant height. The apparent decrease in the carbon dioxide flux, i.e. the rate of photosynthesis, in the wheat crop at about noon is often seen and is usually attributed to a temporary closure of the stomata caused by a greater rate of transpiration than water uptake. The estimated errors in the flux of carbon dioxide were 10 per cent for wheat and 20–30 per cent for pine, and may have been greater because of uncertainties about carbon dioxide movement from the soil, the so-called soil respiration. This latter effect can be seen in Fig. 5.23, which shows the contributions of the soil and three layers of the wheat canopy to the total flux of carbon dioxide. Experiments in England have suggested that the carbon dioxide flux from the soil contributes up to 20 per cent of the carbon dioxide absorbed by the crop (cf. also p. 18).

The uppermost leaves of the wheat crop are obviously the sites of most active photosynthesis. In other species the highest rates of photosynthesis appear to coincide with the layers of the canopy with the greatest leaf area densities (Figs. 5.24 and 5.25). This last set of data was obtained using a completely different technique – the measurement of the rate of uptake of $^{14}CO_2$ which was fed to the leaf for a few seconds.

Analyses of profiles of this type depend on a knowledge of the arrangement of the leaves within the canopy, since this will determine the light profile within the canopy and hence the energy available for photosynthesis (cf. Fig. 2.10). In addition, it is often necessary to consider the area of all the organs exposed to light since most are capable of photosynthesis and often intercept large proportions of the incoming radiation. For example, the surface area of leaf sheaths and stems can comprise up to 50 per cent and 13 per cent of the total photosynthetic areas of wheat and maize, respectively. In addition, the tassels of maize can intercept up to 9 per cent of the daily insolation at commercial crop densities. The photosynthesis of less obviously photosynthetic organs can, in some circumstances, be very significant. The carpels of beans and peas are photosynthetic, as are some woody shoots such as those of grape. Photosynthesis by the ear of barley is sufficient to supply about 45 per cent of the final ear dry weight. In awnless wheat the contribution of ear photosynthesis to the final ear weight is much less than in barley, but the situation in awned wheat is more like that in barley (cf. Table 8.1).

A simple approach to light interception has been given on pp. 23–7. More detailed analyses involve not only the area of the leaves within each layer of the canopy but also their angle relative to the solar radiation. Since the

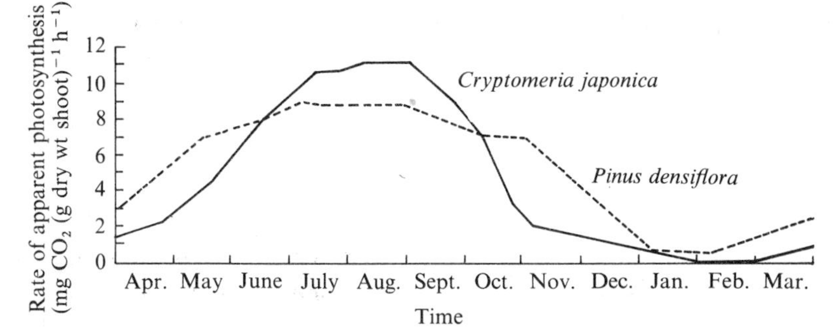

Fig. 5.26. The seasonal changes in net photosynthesis of *Cryptomeria japonica* and *Pinus densiflora*. The measurements were made at 20 °C and saturating irradiance. After Negisi, K. (1966). *Bull. Tokyo Univ. For.* **62**, 1–115.

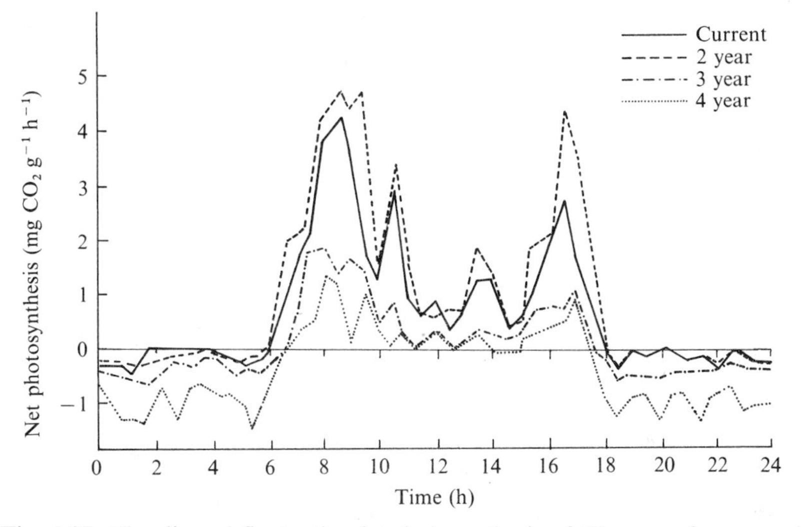

Fig. 5.27. The diurnal fluctuation in photosynthesis of *Pinus ponderosa* needles of four different ages. After Helms, J. A. (1970) *Photosynthetica*, **4**, 243–53.

angle of the sun's rays shows both diurnal and seasonal variations, this type of analysis becomes complicated and is beyond the scope of this book. The situation, found in some plants, where the leaves show diurnal changes in position, has not yet been incorporated into these analyses.

A prediction of models of this type is that leaf angle could be an important determinant of photosynthetic rate and hence crop yield. At low leaf area indices, horizontal leaves intercept more radiation than do more vertical leaves, but the reverse is true at larger leaf area indices. More erect leaves should also be more efficient at low latitudes, i.e. high sun angles, but are less important at higher latitudes. These advantages are probably minor compared with those accruing from good husbandry, and other plant components, but they might become more important under intensive cultivation, and erect leaves are being incorporated into some new cereal varieties.

Seasonal trends

In annual crops seasonal trends in photosynthesis can usually be explained reasonably simply in terms of the radiation available and the effects of other environmental and internal factors. The situation in deciduous perennials is similar to that in annuals, with the additional possibility of some photosynthesis by woody parts after leaf fall. In evergreen perennials the situation becomes more complex in that the environmental variations can be more extreme. Photosynthesis can proceed, however, in some species at very low temperatures, and hence there can be a positive net photosynthesis and dry weight accumulation throughout the year (Fig. 5.26). The analysis of photosynthesis within the canopy is complicated by branches within the canopy receiving very different supplies of mineral nutrients. Moreover, these branches bear leaves differing in age by several years (Fig. 5.27), with the older leaves usually being in the more shaded positions.

RESPIRATION

The amount of growth which can be made by a plant or crop is a function of the net rate of photosynthesis, P_N, i.e. the difference between the rates of gross photosynthesis, P_G ,and respiration, R

$$P_N = P_G - R \qquad 5.10$$

Photosynthesis has been discussed at length above, but respiration is a more difficult issue to put into context. It is unnecessary here to consider the detailed biochemistry of respiration. It should be remembered, however, that it involves the oxidation of complex substances such as sugars and fats. Various intermediates in the respiratory processes are necessary for the synthesis of other materials, and the energy for these syntheses is provided by ATP and NADPH formed during respiration.

The constituents of the plant are constantly turning over and hence some of the respiration is needed to support this and the maintenance of the cellular organization. This fraction has been termed the 'maintenance respiration', R_M. The remaining intermediates and energy can be used to synthesize additional materials, i.e. produce new growth, and this can be called 'growth respiration', R_G, where

$$R = R_M + R_G \qquad 5.11$$

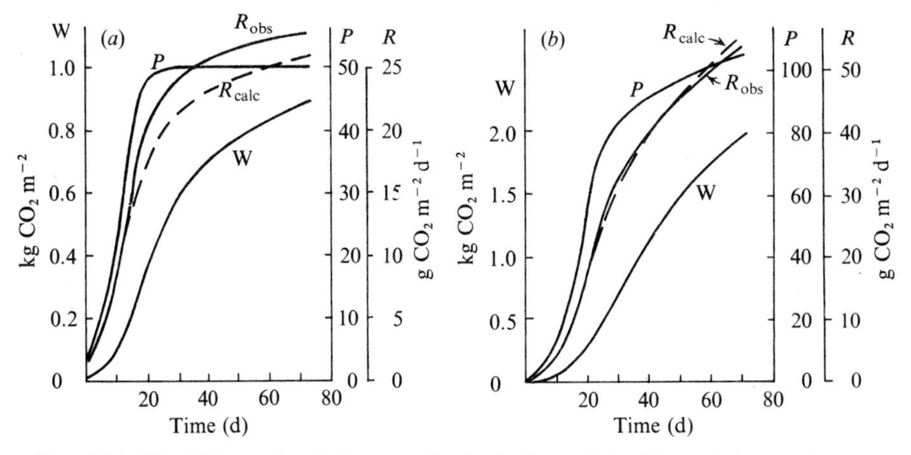

Fig. 5.28. The 24 h totals of photosynthesis, P, dry weight, W, and observed, R_{obs}, and calculated respiration, R_{calc}, of *Trifolium repens* plants grown at 100 W m^{-2} under two different lighting arrangements (*a* and *b*). After McCree, K. J. in Šetlík (1969).

It must be emphasized that these are not two types of respiration. Equation 5.11 describes how the products of a single process are allocated to two purposes, both synthetic in nature.

The maintenance respiration is considered to be a function of the dry weight of the plant or organ, whereas the growth respiration seems to be related to the gross rate of photosynthesis or supply of assimilates. On this basis, the total rate of respiration would be given by

$$R = a\mathrm{W} + bP_{\mathrm{G}} \qquad 5.12$$

where a and b are constants. This equation gives a good prediction of the rate of respiration of small canopies of *Trifolium repens* grown at 20 °C with $a = 0.015$ d^{-1} and $b = 0.25$ (Fig. 5.28). The value of b indicates that the plants respire 25 per cent of the carbon they fix. A slightly different approach using cotton leaves at 30 °C gave values of a and b of 0.026 d^{-1} and 0.54 respectively, and similar values for cotton bolls were 0.003 d^{-1} and 0.38.

The production of energy during growth respiration involves the complete respiration of assimilates, and hence in a time interval, Δt, a loss of dry matter by the plant of ΔS_R. The difference between ΔR_G and $\Delta S_R/t$ is the amount of dry matter available for growth, ΔS_G. Therefore

$$\Delta S_G/\Delta t = R_G - \Delta S_R/\Delta t \qquad 5.13$$

The efficiency of production of new growth, Ω, is given by

$$\Omega = \Delta S_G/(\Delta S_G + \Delta S_R) \qquad 5.14$$

and the overall efficiency of assimilate utilization by the plant, ω, is

$$\omega = \Delta S_G/(\Delta S_G + \Delta S_R + \Delta S_M) \qquad 5.15$$

where
$$\Delta S_M/\Delta t = R_M = mW \qquad 5.16$$

where m is a maintenance coefficient.
From Equation 5.14

$$\Delta S_G/\Delta t = \{\Omega/(1-\Omega)\}\Delta S_R/\Delta t \qquad 5.17$$

and from Equations 5.10, 5.11, 5.13 and 5.16

$$P_G = \Delta S_G/\Delta t + \Delta S_R/\Delta t + \Delta S_M/\Delta t \qquad 5.18$$

and
$$R = \Delta S_M/\Delta t + \Delta S_R/\Delta t \qquad 5.19$$

From 5.17, 5.18 and 5.19

$$R = m\Omega W + (1-\Omega)P_G \qquad 5.20$$

which is the same form as Equation 5.12 and can be arranged to give

$$R = m + \{(1-\Omega)/\Omega\}\{(dW/dt)/(1/W)\} \qquad 5.21$$

from which m and Ω can be estimated if R and the relative growth rate are known. Experiments of this type on cotton bolls have produced values for m and Ω of 0.006 g g^{-1} d^{-1} and 0.74 g dry matter (g substrate)$^{-1}$ respectively; these are in reasonable agreement with the values quoted above.

The continued maintenance of previously formed structures would be expected to take precedence over the production of extra dry matter. If the supply of current assimilates is insufficient for this purpose the organ in question, say a leaf, uses reserve assimilates. If no reserves are available there must be a net breakdown of the structure of the leaf and this would be the start of the senescence of that leaf. A similar sequence of events probably occurs during the growth of a plant. As the plant increases in size it will accumulate an increasingly large proportion of non-photosynthetic, but respiring, tissues. The respiratory load will, therefore, tend to increase faster than the supply of assimilates and, eventually the total dry weight decreases as the reserves are used and senescence occurs.

It is difficult to make unequivocal estimates of the rate of respiration. They are usually dependent on the diffusion of carbon dioxide or oxygen into and out of the tissues and hence are sensitive to the various resistances to gaseous diffusion and the relative rates of respiration and photosynthesis. The rate of carbon dioxide exchange in the dark provides one estimate, but this would be dependent on the rate of photosynthesis in the previous light period and the various diffusive resistances. In addition, in C_3-plants it would provide no estimate of the rate of photorespiration. A useful measure in the estimation of respiration is the flux of carbon dioxide into the inter-

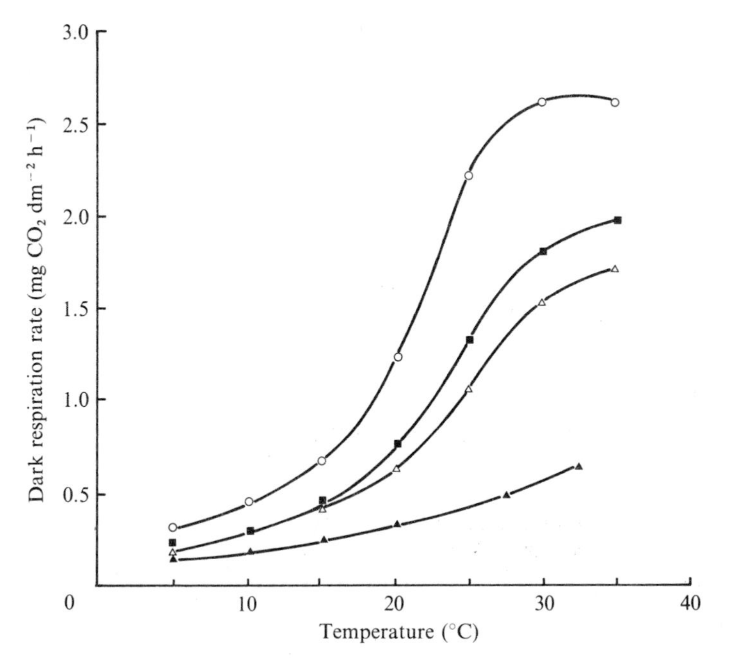

Fig. 5.29. The effect of temperature on the rate of respiration of various leaves of a cotton plant. (circles) Tenth leaf, half expanded; (crosses) sixth leaf; (open triangles) fourth leaf after one day at high light intensity; (closed triangles) fourth leaf after a day in darkness. After Ludwig, L. J., Saeki, T. & Evans, L. T. (1965). *Aust. J. biol. Sci.* **18**, 1103–18.

cellular spaces. This flux will be a function of the concentration of carbon dioxide in the intercellular spaces, C_W, and can be measured in a system where known concentrations of carbon dioxide in air or nitrogen are *forced* through a leaf (cf. Fig. 5.6). The rate of exchange of carbon dioxide, F, is a linear function of C_W such that

$$F = -a + bC_W \qquad 5.22$$

If photorespiration is absent or prevented, b in Equation 5.12 corresponds to $1/r'_m$ and a is the rate of release of carbon dioxide into the intercellular spaces. When there is some photorespiration a includes a component due to this, and it can be estimated from measurements made in the presence and absence of oxygen. An experiment of this type with cotton leaves at 27 °C yielded values of r'_m of 3.0 s cm^{-1} and of the rate of respiration in oxygen-free air of 1.3 mg CO_2 dm^{-2} h^{-1}. Photosynthesizing leaves in air were found to respire 38 per cent of the assimilated carbon dioxide.

The response of respiration to environmental changes is usually different from that of photosynthesis (cf. Fig. 5.12). An important difference between

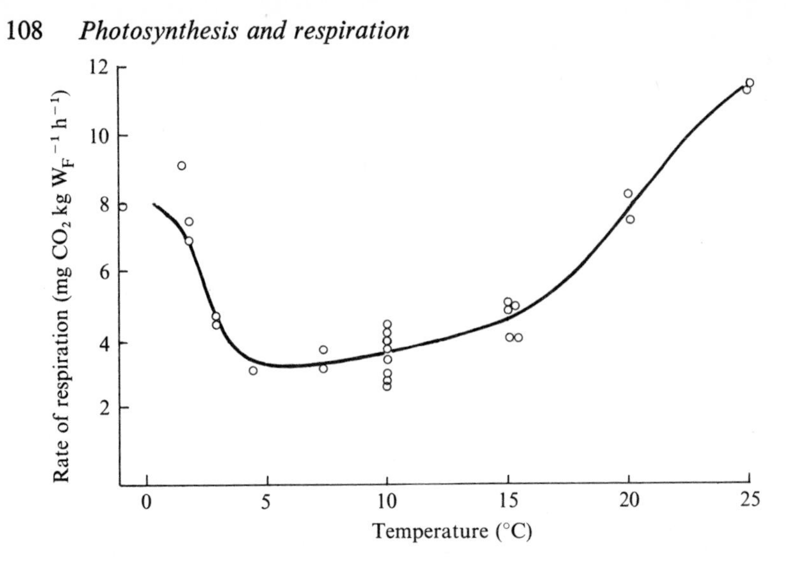

Fig. 5.30. The rate of respiration of potato tubers after storage at a range of temperatures. After Burton, W. G. (1966). *The potato*. Veenman & Zonen, Wageningen.

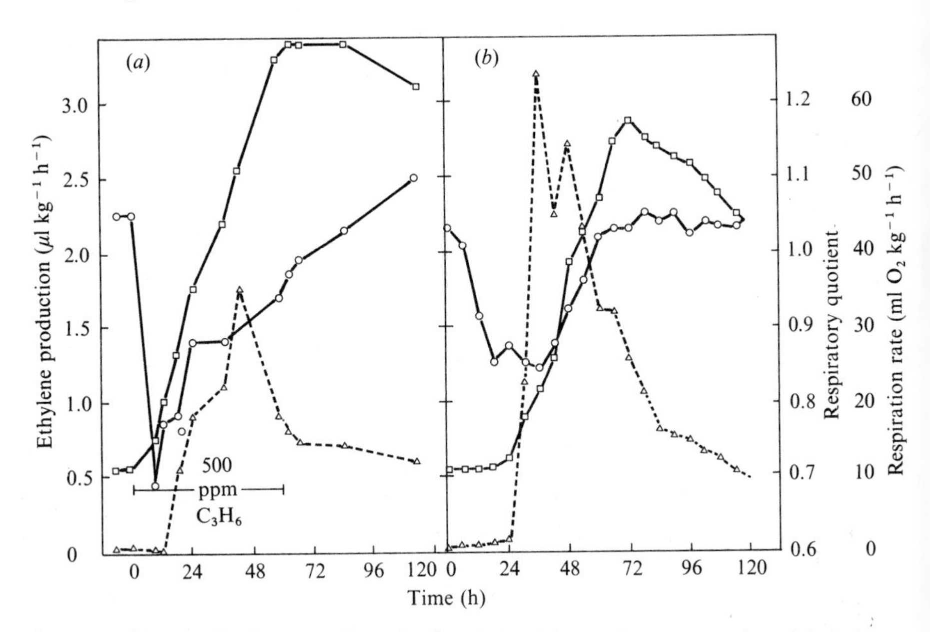

Fig. 5.31. Changes in the rate of respiration (squares), respiratory quotient (circles) and ethylene production (triangles) of bananas. (*a*) Climacteric initiated by exposure to 500 ppm propylene. (*b*) Natural initiation of climacteric. After McMurchie, E. J., McGlasson, W. B. & Eaks, K. (1972). *Nature, Lond.* **237**, 235–6.

the two processes is in their response to temperature. The Q_{10} of respiration is usually about 2 with an optimum in cotton, for example, at about 30 °C (Fig. 5.29). The Q_{10} of photosynthesis is usually less than 2. In some circumstances other responses are possible, and potato tubers which have been stored for a period at 0 °C have a similar respiration rate to others which have been stored at 20 °C because of the increased rate of conversion of starch to sugars at low temperatures (Fig. 5.30). This effect is not seen unless there is sufficient time for the concentration of sugars to increase; measurements made for short periods show the usual temperature response.

There is usually a gradual decrease in the rate of respiration as an organ ages (cf. Fig. 5.18) and some fruits, e.g. citrus, follow the same pattern. Other fruits, however, such as apples and bananas, show a marked increase in the rate of respiration as they ripen (Fig. 5.31). This increase is known as the climacteric and, in addition to changes in the respiration rate and respiratory quotient, there are associated changes in the composition, colour and texture of the fruit. The stimulus causing the onset of the climacteric is thought to be the production of ethylene, and as respiration increases yet more ethylene is produced. If the initial stimulus is replaced by exposure to propylene the subsequent progress appears to be quite normal.

FURTHER READING

Eckardt, F. E. (ed.) (1968). *Functioning of terrestrial ecosystems at the primary production level.* UNESCO, Paris.

Hatch, M. D. & Slack, C. R. (1970). Photosynthetic CO_2-fixation pathways. *Ann. Rev. Pl. Physiol.* **21**, 141–62.

Hatch, M. D., Osmond, C. B. & Slatyer, R. O. (ed.) (1971). *Photosynthesis and photorespiration.* Wiley-Interscience, New York.

Heath, O. V. S. (1969). *The physiological aspects of photosynthesis.* Heinemann, London.

Hesketh, J. D., Baker, D. N. & Duncan, W. G. (1971). Simulation of growth and yield in cotton: respiration and the carbon balance. *Crop Sci.* **11**, 394–8.

Jackson, W. A. & Volk, R. J. (1970). Photorespiration. *Ann. Rev. Pl. Physiol.* **21**, 385–432.

Šetlík, I. (ed.) (1969). *Prediction and measurement of photosynthetic productivity.* Proc. IBP/PP Tech. Meeting, Trebon. PUDOC, Wageningen.

Thornley, J. H. M. (1970). Respiration, growth and maintenance in plants. *Nature, Lond.*, **227**, 304–5.

Zelitch, I. (1971). *Photosynthesis, photorespiration and plant productivity.* Academic Press, New York and London.

6

Germination and Seedling Emergence

DORMANCY

Most seeds, during their development on the mother plant, pass from a state in which their embryos are able to continue growth and development if placed in suitable conditions to one in which further growth will not take place. This dormant state may extend, depending on species, from a few days to several years. Growth will not be resumed, no matter how favourable the environment, until the seed (and, indeed, also vegetative propagules) has passed through this state. Current theories envisage dormancy as resulting either from an excess of growth-inhibiting over stimulating substances or from the presence of water-impermeable seed coats, as in seeds of many Leguminoseae. Abscisic acid is widely favoured as the principal inhibitor in many species; however, it is not the active agent in *Sinapis* or in *Corylus* and there are probably a large number of endogenous inhibitors. Gibberellins and ethylene are often credited with being the active promotive agents although treatment with other substances will also break dormancy. In natural environments, the factors leading to loss of dormancy vary widely with species; these include prolonged exposure to low temperature, periods of alternating temperature, leaching of water-soluble substances, weathering of seed coats, gradual loss of the capacity to produce inhibitors, and so on.

Limitations of space do not permit us to explore this feature. It is usually not significant in the production of the major annual crop plants, the seeds having passed through this state during storage between harvesting and sowing. However, it is important with seeds of some pasture species, ornamentals, forest species, weed species and annual species where the period between harvest and planting is very short. It is also a prominent feature of tubers, bulbs and corms, and the aerial buds of deciduous tree fruits.

PROCESSES DURING GERMINATION

Water uptake

The only requirements for a non-dormant seed to resume active growth and development are the uptake of water at a non-restrictive temperature and sufficient oxygen. When a seed is placed in a moist medium, water uptake

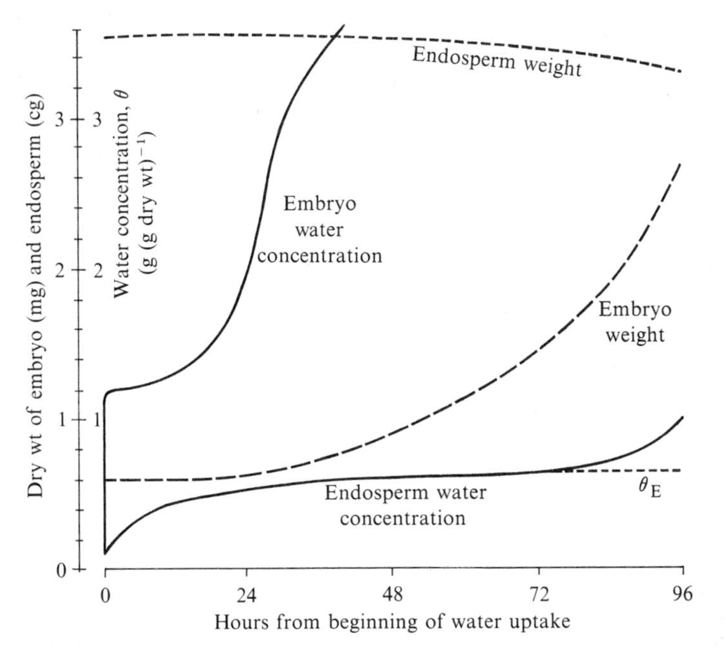

Fig. 6.1. Changes in water concentration (continuous lines) and in weight (broken lines) of embryo and endosperm of wheat. Note that scale of embryo weight is 10 times that of endosperm weight (i.e. initially the embryo is 1/60th grain weight). θ_E is the equilibrium water content of the endosperm as defined in Equation 6.1. Data after Milthorpe, F. L. (1950). *Ann. Bot.* **14**, 79–89, with further unpublished measurements.

may commence immediately or there may be a lag period of up to several hours before there is any perceptible uptake. This lag phase may vary widely between different individual seeds of the same sample – and, hence, is rarely detected if measurements are made on samples of many seeds. It appears to become pronounced if the moisture content is less than 0.10–0.12 g (g dry weight)$^{-1}$ and to disappear if the seed is conditioned to higher moisture contents before sowing. Although in most legumes water enters only through the strophiole, in others it enters only via the hilar fissure; the rate of entry is then associated with the changing width of the opening related to hygroscopic adsorption. However, this effect of initial moisture content is also found in some species which absorb water through all parts of the seed coat.

The course of water uptake can be most readily followed in a cereal grain in which the growing embryo is spatially separated from the storage endosperm (Fig. 6.1). There is first a very rapid uptake by the embryo, the water concentration changing from the air-equilibrium value of 8–12 per cent to about 120 per cent (on a dry weight basis). This presumably represents

Table 6.1. Diffusivity, radii and moisture characteristics of three species of seed germinated in aerated distilled water at 28 °C. Germination was assumed to be complete when the radicle was 2–4 mm long. After Phillips, R. E. (1968). *Agron. J.* **60**, 568–71.

Species	t_g(h)	θ_g(mg seed^{-1})	R at $\theta_{g/2}$(mm)	D (10^{-4}cm^{2}h^{-1})
Cotton	62	133	0.32	2
Maize	54	129	4.58	8
Soybean	20	214	4.19	32

uptake by the pre-existing cells to full turgor. The metabolic activity of the embryo is then accelerated. The water concentration of the endosperm increases much more slowly. Initially, this is primarily by imbibition, the water flowing in response to the difference in water potential (about 10^5 J kg^{-1} between air-dry seeds and free water). Proteins and cell-wall materials, but apparently not starch, swell on imbibition, the volume increasing by some 40 per cent during the first 24–30 hours. (This is comparable to the decrease in volume which occurs on ripening.) Where the water content of the whole grain – or seed – is followed, the changes are, of course, dominated by those in the endosperm – or storage regions – simply because it is a much greater proportion of the total weight than the embryo. During this initial phase (say, up to about 60 h in Fig. 6.1), the water uptake can be described by

$$\theta = \theta_E - (\theta_E - \theta_0)\exp(-at) \qquad 6.1$$

where θ_0 and θ_E are the initial and equilibrium moisture contents, of the endosperm, respectively, and a may be regarded as a diffusion coefficient reflecting transport within the seed divided by some proportion of the surface area of the seed (such as that permeable to water, amount in contact with water in the medium, etc.). The water potential of the seed will be related to its moisture content by a curve similar in form (but not in position) to that of Fig. 2.3 (p. 16). A more explicit description of the diffusion coefficient, D, for water in seeds can be obtained from an expression involving infinite series:

$$\frac{\theta}{\theta_g} = \frac{\pi^2/6 - \sum_{n=1}^{\infty} (1/n)^2 \exp(-Dn^2\pi^2 t/R^2)}{\pi^2/6 - \sum_{n=1}^{\infty} (1/n)^2 \exp(-Dn^2\pi^2 t_g/R^2)} \qquad 6.2$$

where θ_g and t_g are the water content and time respectively at some defined stage of germination, and R is the average radius of the seed, assumed to be spherical and conveniently taken as that existing when half of the water required for germination has been absorbed. (D needs to be evaluated graphically; cf. J. Crank, *The mathematics of diffusion*, Oxford University Press, 1956, Section 6.3.) Values for three different species showed a three-fold range in time to germination, a less than two-fold range in the water content at germination and about a sixteen-fold range in diffusivity (Table 6.1).

Gas exchange

In an air-dry seed, the rate of gas exchange is extremely low. However, with the uptake of water it increases rapidly, the rate of carbon dioxide output often approximating to an exponential function of the water content during the early stages. Until the testa has been ruptured, it constitutes a barrier to the diffusion of oxygen into the seed. For example, the coefficients for the diffusion of oxygen at 25 °C through the testa and embryo of seeds are about 10^{-7} and 10^{-2} $mm^2\ s^{-1}$ respectively, compared with 10^{-3} for water. It seems that in most seeds oxygen is in short supply and that often fermentation is an important, if not the main, respiratory pathway until the seed coats are ruptured. Removal of the seed coats often leads to a doubling or trebling of the rate of oxygen uptake. However, some species, such as rice, are able to germinate in the absence of oxygen, whereas wheat fails completely.

Features of metabolism

In cereals, one of the earliest detectable features of the increased metabolic activity is the activation of pre-existing messenger-ribonucleic acid accompanied by an increase in the capacity of the embryos to synthesize protein; this follows the imbibition of water by only a few minutes. Polysomes (organelles, composed of ribosomes and messenger-ribonucleic acid, which incorporate amino acids into protein) soon begin to increase, as does the rate of respiration and the extension of some cells (especially, in cereals, those of the coleorhiza, the extension of which leads to the rupture of the pericarp). There is very little synthesis of new deoxyribo- or ribonucleic acids within the first 12–16 hours. By this time there has been some extension growth of existing cells and the early stages of mitosis can be detected. During the first 24 hours or so, the extension and division growth depends entirely on the use of amino acids, fats and soluble carbohydrates stored in the embryo. Also, during this period, a significant amount of gibberellin is secreted by the embryo, possibly from the scutellar region. This diffuses to the aleurone cells and stimulates them to synthesize and release hydrolytic enzymes, particularly α-amylase and proteases, into the endosperm. These lead ultimately to the production of glucose, fructose and maltose, which are transferred through the scutellum, where sucrose is synthesized, to the embryo. Most of the nitrogen is probably transferred as glutamine.

Analogous changes of water, substrates and metabolism occur in the

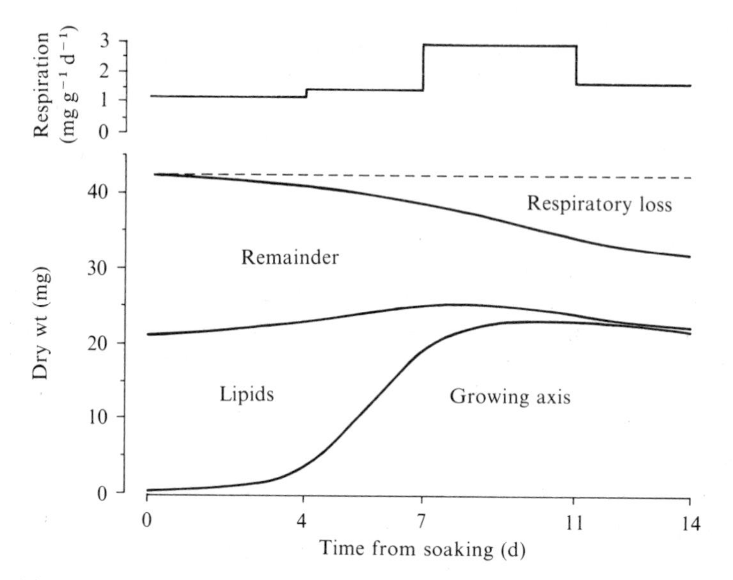

Fig. 6.2. Changes in weight of growing axis and cotyledons (lipids and remainder), together with respiratory losses, of *Citrullus vulgaris* seedlings growing in the dark. The average respiration rates are shown in the upper graph. Data after Hardman, E. E. & Crombie, W. M. (1958). *J. exp. Bot.* **9**, 239–46.

growing axes and the cotyledonary storage regions of dicotyledonous seeds, although these are complicated by the spatial relationships found. However, current evidence suggests that gibberellins do not play so prominent a role in mobilizing the reserve materials and that other, unknown, factors are involved. Studies with peas (hypogeal, non-growing cotyledons) have shown that the growing axis is not necessary for the hydrolysis of starch and protein reserves in the cotyledons but is required for the elaboration of a complex sub-cellular structure by these organs. This structure consists of well-developed mitochondria, an elaborate smooth endoplasmic reticulum, extensive vesicles and very lobed nuclei (indicating active metabolism). Transfer of materials from the cotyledons to the growing axis does not occur until this structure develops, the axis using its own reserves until the radicle is about 4 cm long.

Examples of the overall changes in transfer between reserve and growing regions of a species with epigeal germination are shown in Fig. 6.2. By Day 11, before which time the plant would normally become autotrophic, some 25 per cent of the translocated material had been used in respiration. This seems to be about the usual proportion, one-quarter to one-third of the seed weight being dissipated in establishing the plant. Almost all of the readily mobilized substances, such as starch, sugars, lipids and proteins, are moved to the growing plant, irrespective of environment; little more than a dilute solution of sugar and cellulose remains in the storage regions of those seeds in which the storage tissues do not grow. (The cell walls of endosperms of most cereals consist mainly of hemicelluloses, which are readily broken down during germination.) However, the cotyledons of those plants with epigeal germination, such as *Citrullus* and other cucurbits, normally grow and function as foliage leaves; they have very high rates of photosynthesis.

The emergence of the coleoptile or the plumule above the ground brings the plant under the influence of light and results in the suppression of mesocotyl, hypocotyl or epicotyl growth and stimulates the formation of chlorophyll. There follows a transition period during which photosynthesis gradually assumes pre-eminence over seed reserves in supplying the substrates for growth. Seedlings of wheat become completely independent about the stage when the second leaf has fully emerged and the third is just emerging. Cucumber seedlings become independent soon after unfolding of the cotyledons. However, seedlings are dependent on supplies of nitrogen, phosphorus, zinc and possibly most minerals from the soil during emergence in order to achieve maximum rates of growth. The reserves laid down in the seed usually have a preponderance of carbohydrates or lipids over the major mineral elements in respect of the proportions required for the early growth of the seedling. This is offset to some extent by the early growth being predominantly that of the roots, more than three times the amount of reserves being diverted to root growth than to leaf growth. In other words, the seedling establishes independence for supplies of minerals and water some time before photosynthesis is fully operational. (This may not be true of all seeds, especially those such as *Salsola*, in which the shoot grows first.)

RESUMPTION OF BUD GROWTH

Vegetative propagules, such as tubers, bulbs and corms, and the buds of perennial plants, also usually pass through a period of dormancy. The buds of most deciduous tree fruits require a long period of exposure to low temperatures (below 3–4 °C) before they will resume growth with the increasing temperatures in spring. No environmental agent appears to influence dormancy of buds of the potato tuber; this state is gradually lost with time (in inverse relation to the age and position of the bud). It must be associated with some internal state; for example, there appears to be a decrease in concentration of abscisic acid and an increase in that of gibberellins. However, the evidence is not clear and unequivocal. The metabolism of the tuber tissue appears to change from a predominance of synthesis and deposition during its growth and storage to a marked increase in hydrolytic reactions at the end of dormancy.

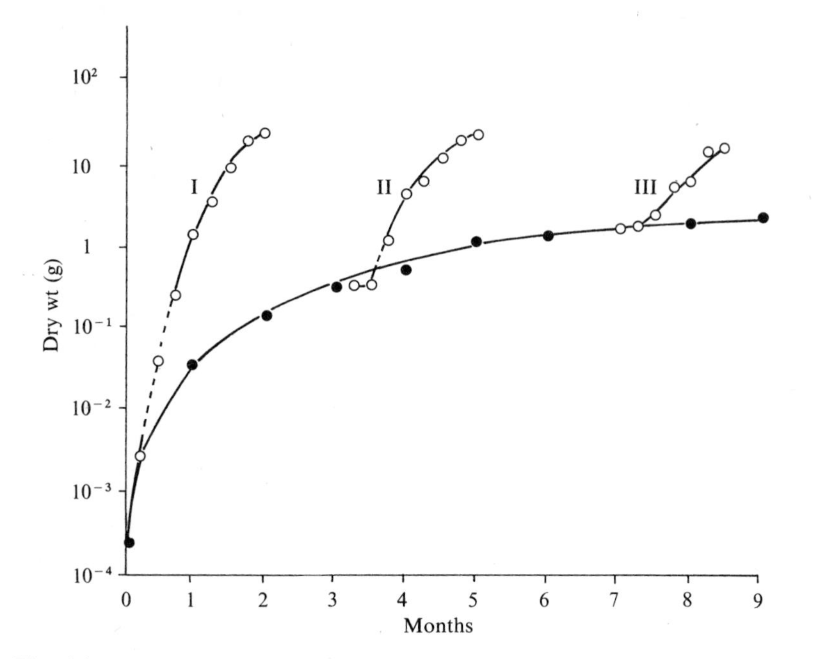

Fig. 6.3. Increase in dry weight of sprouts on a potato tuber during storage (closed circles) and on transfer to moist soil (open circles) at three stages. The broken lines show growth prior to emergence. After Headford, D. W. R. (1962), *Eur. Potato J.* **5**, 14–22.

Where tubers continue to be maintained in storage, the apical bud, and then the lateral buds in basipetal succession, elongate at rates dependent on temperature. The rate of growth also depends on the total reserve of substrate available in the tuber and the rate at which this is hydrolysed, a process in turn dependent on the transmission of some stimulus from the growing sprout. Although it has not been shown that this stimulus is a gibberellin, externally supplied gibberellic acid will initiate all the changes observed. Soon some of the sprouts – the latest to start growing and hence the smallest – stop growing due to inhibition by apical-dominance phenomena. (These will start growing again if isolated from the growing sprouts.) There is also competition for substrates between those sprouts which continue to grow. This competition is almost certainly for inorganic mineral nutrients as the degree of interference can be reduced, and even eliminated, by supplying inorganic nutrients.

During storage, growth proceeds at a relatively slow rate. On provision of liquid water following planting, this rate greatly accelerates (Fig. 6.3), but the relative growth rate is always less, the greater the size of the sprout(s) at planting. Mineral nutrients are absorbed from the external medium and reserves from the mother tuber continue to be transferred to the growing shoot until they are effectively exhausted. For example, in one experiment, it was found that by the time a new shoot had produced 300 cm^2 leaf area, photosynthesis had contributed only 7 per cent of the dry weight. By this time, the mother tuber was little more than a dilute solution of sugar held within semi-permeable membranes and cell walls.

THE INFLUENCE OF THE ENVIRONMENT

Evaluating the performance

Only three components of the environment – temperature, water supply and depth of sowing – are usually significant in the germination performance of non-dormant seeds, although light is occasionally important. In considering these, we need also to use parameters which allow a useful and quantitative evaluation. The 'germinability' of a sample of seed is defined as the proportion of the number sown which reaches a given stage of growth. This is frequently taken as the emergence of the radicle – a stage dominated by imbibition and cell extension – or it may be extended to emergence of the shoot from the soil and the establishment of complete autotrophy. (Some authors distinguish between the stages of *germination* ending with radicle emergence from the seed and *pre-emergence* growth ending with some morphological index of shoot emergence above the soil.) This time distribution of the number of individuals reaching the defined stage is often of the form shown in Fig. 6.4; it can be described approximately by

$$p = A[1-\exp\{-k(t-t_0)\}] \qquad 6.3$$

where p is the proportion germinated at time t, t_0 the time to germination of the first seed to germinate, A the maximum proportion to germinate (the *germination percentage* or *germinability*), and k measures the spread in the population of the time to germination. The same shape of curve is usually found even when some state of emergence is taken as the defined stage. The rate at which germination progresses is given by the reciprocal of the time required; i.e. the rate of germination of the fastest germinating seed is given by $1/t_0$ and that of the median seed by $k/(kt_0+0.693)$. (The time when $p = A/2$, i.e. when half of the seeds capable of germinating have germinated, is given by $t_0+0.693/k$.) Equation 6.3 implies that the number germinating each successive hour falls continuously from t_0, i.e. $dp/dt = k(A-p)$. That this is not exactly so is shown in Fig. 6.4; nevertheless, the fit is very close over most of the range. It might be expected that the rates of germination would be normally distributed; if so, the curves relating p to t would be asymmetrically sigmoid but with the point of inflection more displaced to the right than found here or in most germination data.

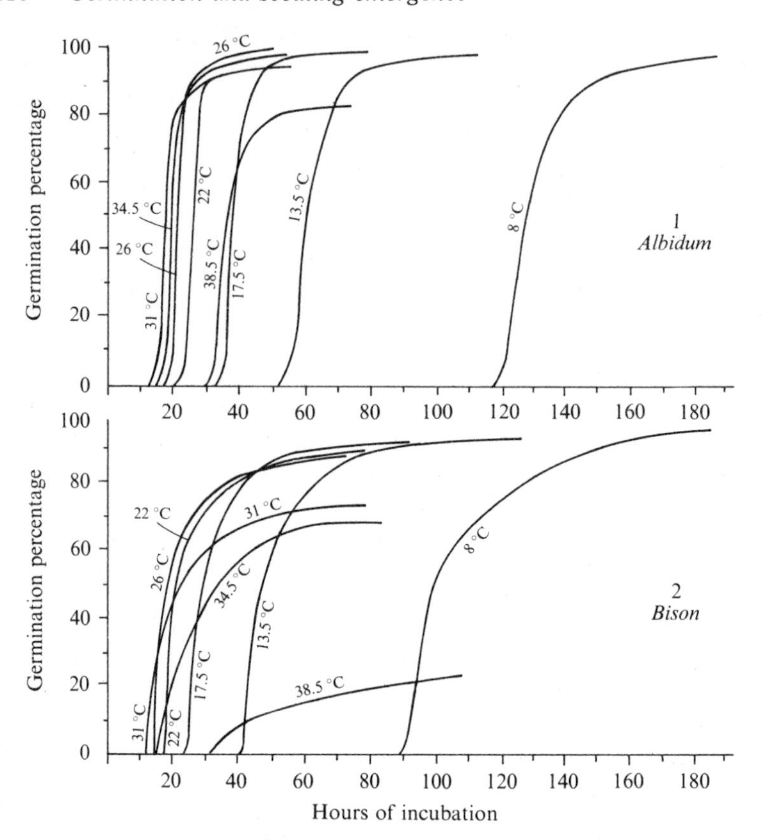

Fig. 6.4. Time distribution of germination (radicle 2–4 mm long) of two cultivars of flax seeds at different temperatures. After Veerhoff, O. (1940). *Am. J. Bot.* **27**, 225–31.

In other words, usually those seeds with the fastest rates of germination comprise the larger proportion of the population.

Temperature

The values of the parameters of Equation 6.3 have been extracted for the data of one of the varieties shown in Fig. 6.4 (Table 6.2). These showed a progressive decrease in A throughout with increase in temperature and a maximum value of k – which measures the degree of steepness of the curves shown in Fig. 6.3 – around 17.5–22 °C. The rate of germination increased up to a temperature of about 32 °C, if the first seed to germinate was taken, or about 28 °C if the median seed was taken. The decrease in rate at high temperatures and the progressive downwards shift of this rate with time is a usual response; the curve is regarded as a resultant between

Table 6.2. Values of the parameters of the relation $p = A\{1-\exp[-k(t-t_0)]\}$ for the data of Fig. 6.4. The rates of germination of the first and the median seeds to germinate are denoted by $1/t_0$ and $1/t_{A/2}$ respectively.

Temperature (°C)	t_0 (h)	k (h^{-1})	A (%)	$1/t_0$ (h^{-1})	$1/t_{A/2}$ (h^{-1})
8	81	0.042	96	0.0123	0.0103
13.5	38	0.080	93	0.0263	0.0215
17.5	21	0.115	92	0.0571	0.0366
22	17	0.113	90	0.0595	0.0437
26	12	0.097	88	0.0833	0.0524
31	9	0.070	80	0.1111	0.0526
34.5	14	0.071	68	0.0714	0.0417
38.5	26	0.032	23	0.0384	0.0210

Table 6.3. Minimum, optimum and maximum temperatures for germination (radicle emergence) of certain crops. Taken from various sources in the older literature.

Species	Minimum (°C)	Optimum (°C)	Maximum (°C)
Lucerne	1	30	38
Rye	2	25	35
Wheat	4	25	32
Maize	9	33	42
Rice	11	32	38
Melon	14	35	40

the increasing rate of enzyme reactions with increase in temperature and an increasing inactivation of the enzymes present. Here, the minimum, optimum and maximum temperatures would be about 4, 28 and 40 °C, respectively. Some data taken from the older literature are shown in Table 6.3. Although these, especially the maxima, must be regarded as being approximate, they do give some idea of the variation found between different species. For example, although spring cereals can be sown when the daily mean soil temperature reaches about 4 °C, maize rarely germinates until the temperature is about 15 °C. The maxima would probably be lower if time to emergence were taken as the criterion; the optima also seem high for some of the lower-temperature species.

A set of observations on periodic sowings of maize made throughout the year shows how the ultimate germination is related to soil temperature (Fig. 6.5). It will be noted that under unfavourable conditions the percentage emergence is also related to rate of germination and growth. This linear relationship was present only when the germination rate of the median

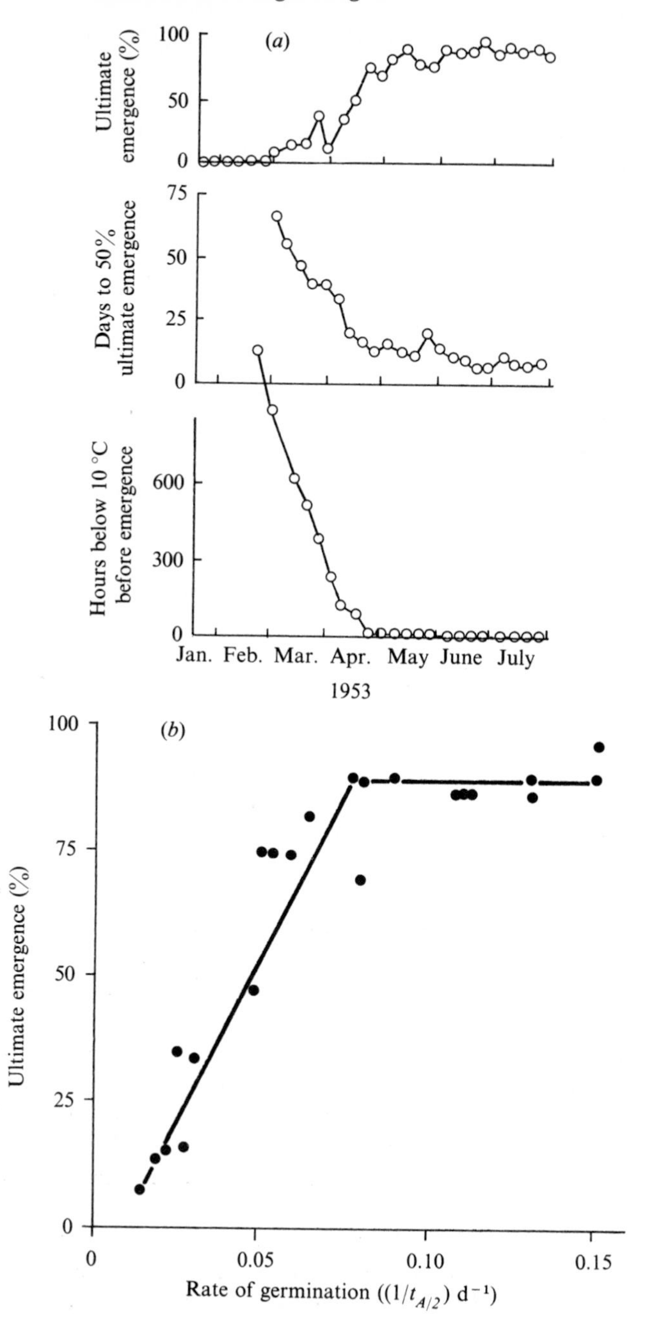

Fig. 6.5. The influence of planting date on the emergence of maize, the time to emergence and the soil temperature at 5 cm in the field at Oxford, England. (*a*) After Harper, J. L., Landragin, P. A. & Ludwig, J. W. (1955). *New Phytol.* **54**, 107–18, and (*b*) the relationship in these data between emergence and rate of germination (taken as the reciprocal of the time when $p = A/2$).

Table 6.4. Time to germination and moisture contents at germination of three species over a range of soil moisture contents. After Phillips, R. E. (1968). *Agron. J.* **60**, 567–71.

Soil moisture content (g g^{-1}) Soil water potential (J kg^{-1})	4.0 -1500	7.5 -300	10.0 -100	20.0 -33
Time to germination (h)				
Cotton	—	72	66	50
Maize	100	58	52	46
Soybean	95	52	42	30
Moisture content of seed at germination θ_g (mg)				
Cotton	—	88	88	88
Maize	100	100	100	100
Soybean	150	150	150	175

seed was less than about 0.08 day^{-1}; that is, when it took more than 12–13 days for the seed to emerge. The death at the lower temperatures was almost certainly related to attack by quasi-parasitic soil fungi and differs, in this respect, from data obtained in laboratory experiments (cf. Table 6.2).

Soil water supply

The relationships of germinability and rate of germination to soil water potential are less clear, mainly because of the difficulty of holding the water potential in the soil at prescribed values. Where it is controlled by contact with solutions of given osmotic potential, the rate of germination is usually reduced over the entire range between 0 and -1500 J kg^{-1} and is completely inhibited around the latter value. However, complications arise because of absorption of the solute (even of mannitol or polyethylene glycol). Where the soil water potential is controlled by the vapour pressure, similar results are found with up to 20 per cent of the seeds germinating at a potential of -3000 J kg^{-1}. Where the matric potential is controlled directly, detectable effects are observed at potentials *above* those often found in soil at field capacity, and there is a reduction both in rate of germination and final germination percentage as the potential is reduced. Much of the effect of the changes in matric potential is attributed to the associated increases in the mechanical strength of the soil matrix rather than to the free energy of the water or to changes in area of contact with liquid water.

It has been shown frequently that seeds will germinate in soil with a water potential of -1500 J kg^{-1} but the rate of germination is slow. Observations in one experiment in which soil was pre-equilibrated to a range of moisture contents and kept at a closely controlled temperature showed that the rate of germination decreased with decrease in water potential over the range from near field capacity to wilting point (Table 6.4). Compare these data with those of Table 6.1. The diffusivities at all soil moisture contents were appreciably less than in distilled water but were still two orders of magnitude less than that of water in soil near wilting point. Moreover, only about 18 per cent of the water in the volume of soil of 1 cm radius around the seed at a moisture content of 4 per cent would need to be absorbed for the seed to germinate. Therefore, although germination is delayed by decrease in water over the available range, it seems likely that seeds of many plants can germinate in soils with a water potential of -1500 J kg^{-1}. However, as with low temperature, attack by semi-parasitic fungi may become more severe under these conditions.

Depth of sowing

The greater the depth of sowing the less the danger that the soil will dry out and the more uniform the temperature (cf. Chapter 2). However, the amount of growth needed to be made by the hypocotyl – or mesocotyl – increases and the greater is the diversion of reserve substrates to this growth. Experiments with *Trifolium subterraneum*, a species with epigeal germination, have shown that the dry weights, but not the areas, of cotyledons at emergence were less, the greater the depth of sowing and that large seed could emerge from a greater depth than small seed. However, subsequent growth was related to cotyledon area (or possibly size of the apical bud) rather than to weight, and hence was entirely a function of the size of seed at sowing.

Germination assessment and seed analyses

The assessment of the ability of commercial samples of seeds to germinate was possibly the first practice to be widely adopted in crop physiology as a routine procedure. This arose, in Europe, mainly with the aim of comparing the value and fixing prices for different seed lots and, in the United States, of advising on the plant-producing ability of the seed. These two somewhat conflicting aims – i.e. reproducibility of tests and of forecasting performance – have now been reconciled by very rigid international rules and procedures (cf. *Proc. Int. Seed Test. Ass.* (1959) **24**, 475). An attempt is made to evaluate any given sample in terms of the proportion which, under strictly defined conditions, is likely to emerge from the soil and sustain autotrophic growth. Abnormal seedlings are noted as well as the degree of dormancy; in addition, the freedom from the presence of other species is assessed. High-quality seed should also be accompanied by a certificate of trueness-to-type and freedom from seed-borne diseases.

FURTHER READING

Borriss, H. (ed.) (1967). *Physiology, ecology and biochemistry of germination.* Ernst-Moritz-Arndt Universität, Greiswald.

Heydecker, W. (ed.) (1973). *Seed ecology*, Proc. 19th Easter Sch. agric. Sci., Univ. Nottingham. Butterworths, London.

Kozlowski, T. T. (ed.) (1972). *Seed biology*, vols. 1–3. Academic Press, New York and London.

Mayer, A. M. & Poljakoff-Mayber, A. (1963). *The germination of seeds.* Pergamon Press, Oxford.

Woolhouse, H. W. (ed.) (1969). *Dormancy and survival. Symp. Soc. exp. Biol.* **23**. Cambridge University Press.

7

Vegetative Growth

'Vegetative growth' is a convenient overall term to describe all those activities associated with the generation and expansion of leaves, the formation of lateral terminal meristems and the growth of some of these as branches, and the concurrent expansion of the root system. Of course, some of these events proceed during the differentiation of the embryo on the mother plant, are resumed during germination and become prominent with the assumption of autotrophic growth. Vegetative growth in any one axis is terminated by flowering: if the apical meristems of all the major stem axes flower at the same time and lateral meristems do not resume growth, vegetative growth of the whole plant also ceases. Such plants are known as monocarpic determinate plants. In some species, such as tomato and *Capsicum*, a lateral bud resumes growth when the terminal meristem is induced to flower and hence phases of vegetative and fruit growth are interspersed in time. In others, only some of the terminal meristems of higher-order axes (e.g. axillary buds) flower and, hence, vegetative and fruit growth proceed concurrently.

GENERATION OF THE LEAF SURFACE

Leaf production

A leaf arises as a lateral appendage of the stem when a group of cells in the outer two layers a small distance back from the terminal meristem commences to divide periclinally and at a much faster rate than the corpus and dome tissues. The primordium usually arises in a position furthest from all existing meristems, suggesting that the latter exert, over some distance, an inhibitory effect on the development of new meristems (Fig. 7.1*a*). After some 3–6 plastochrons, i.e. the time intervals between successive appearances of leaf primordia, a further group of cells commences to divide; these form an axillary bud which is a complete duplication of the entire apex. The cells involved, at least in wheat, are those of the two outer layers (tunica) of the parent axis together with those of the third layer (the sub-hypodermis) from which the core of cells of the bud are derived. The growth of the bud is usually arrested after a short period.

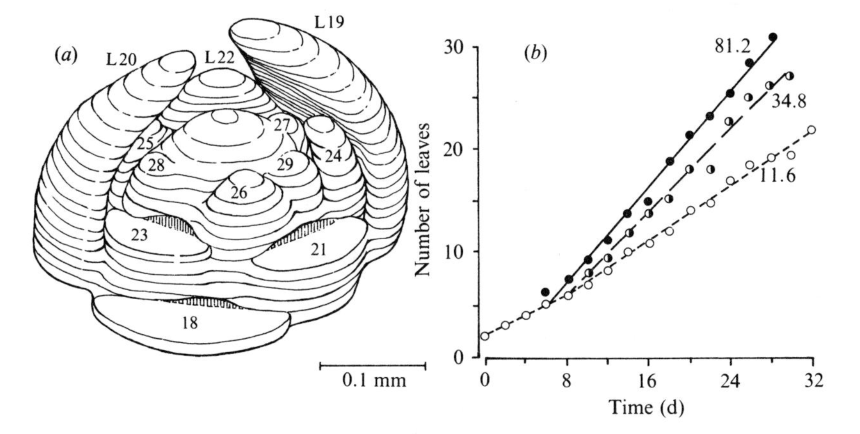

Fig. 7.1. (*a*) Perspective drawing of a shoot apex of an 11-day-old seedling of flax (*Linum usitatissimum* L.). The projection is built up from outline drawings of successive 10-μm sections. Leaves 18, 21 and 23 have been removed to allow the apex and younger primordia to be seen. After Williams, R. F. (1970). *Aust. J. Bot.* **18**, 167–73.

(*b*) Changes in total number of leaves produced with time by cucumber plants kept in darkness until Day 4 and then exposed to the irradiances shown (W m^{-2}) for 15 h daily. After Milthorpe, F. L. & Newton, P. (1963). *J. exp. Bot.* **14**, 483–95.

In any environment in which the irradiance and temperature are maintained constant, the rate of leaf production, dN/dt leaves d^{-1} (and, hence, its reciprocal – the plastochron), by any one stem axis remains constant with time (Fig. 7.1*b*). It exhibits an asymptotic relationship with irradiance, Q_V W m^{-2} d^{-1} – the data in Fig. 7.1*b* being reasonably well represented by $dN/dt = 0.50 + Q_V/(0.72Q_V + 40)$. There is usually an 'optimum' relationship with temperature, the minimum, optimum and maximum temperatures varying with species. Usually, the optimum temperature for cell division – which determines leaf production from an apex as well as the early growth of the leaf primordium – is perhaps 5–6 °C higher than for cell expansion.

Care should be taken to distinguish between the *rate of leaf production* and the *rate of appearance of unfolded leaves.* The former is obtained by counting at constant intervals all leaves on the stem down to the smallest size visible under a stereoscopic microscope, or by assuming that the leaf primordium is initiated when it reaches some arbitrarily defined size (say, 10^{-3} mm^3 or about 500 cells) and estimating this by extrapolation from periodic measurements during growth of the primordium. Rates based on leaves unfolding from the terminal bud (dicotyledons) or emerging from enclosing sheaths (Gramineae) are usually lower than those based on initiation of primordia because leaves require longer than one plastochron

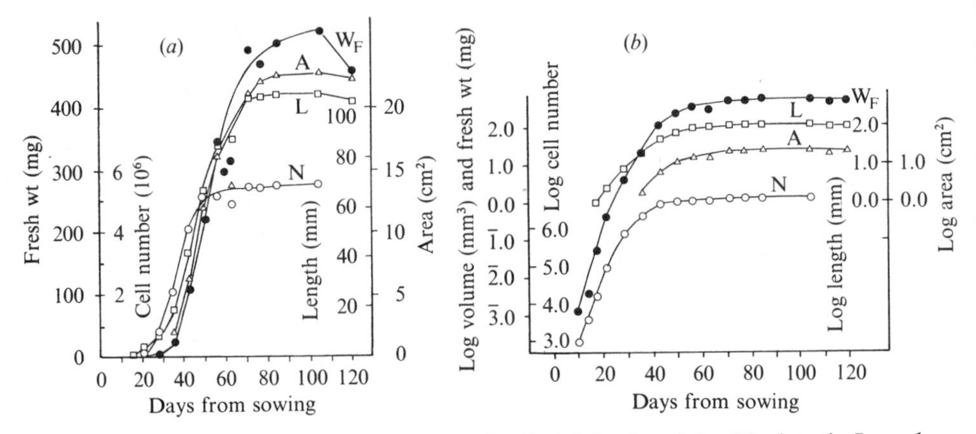

Fig. 7.2. Changes with time in number of cells, N, fresh weight, W_F, length, L, and area, A, of each of the second pair of leaves of sunflower. (*a*) Arithmetic scale and (*b*) logarithmic scale. After Sunderland, N. (1960). *J. exp. Bot.* **11**, 68–80.

to unfold. The number of leaves in the terminal bud therefore increases with time until flower initiation.

The rate of leaf production by the plant as a whole will be determined by the rate of branch production as well as by the differentiation of leaves from each apex. In grasses, especially, leaf production is dominated by the rate of tillering (cf. p. 133).

The growth of a leaf

A considerable number of divisions have no doubt occurred by the time the primordium is first visible under a microscope. The first phase is predominately one of cell division which, in the earlier-formed dicotyledonous leaves in a constant environment, proceeds at an approximately constant relative rate until the time of unfolding from the terminal bud (Fig. 7.2). However, there may be a three- to four-fold increase in average cell volume during this time (mainly in cells of the vascular tissue) and the percentage of cells in mitosis decreases continuously, suggesting that a decreasing proportion of cells is dividing but that those proceeding through division do so at an increasing rate.

After unfolding, cell division continues but at a declining average rate until the leaf is from 0.25 to 0.75 its final size. There is also a surge of cell expansion at this time which, together with cell division, contributes to the rapid increase in the size of the leaf; in the leaf described in Fig. 7.2, for example, the average cell volume increased from about 10^{-6} at initiation to 4×10^{-6} at unfolding to 90×10^{-6} mm^3 at maturity. The leaf reaches its

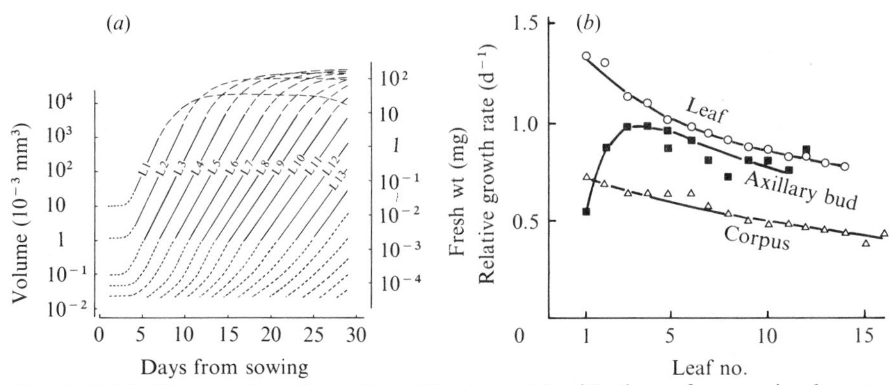

Fig. 7.3. (*a*) Changes in volume (logarithmic scale) with time of successive leaves differentiated from the main axis of *Trifolium subterraneum*. (*b*) The relative rates of increase in volume of the leaves shown in (*a*) during the constant phase of growth together with those of the associated corpus tissues. After Williams, R. F. & Bouma, D. (1970). *Aust. J. Bot.* **18**, 127–48.

final size as the last-formed cells cease expansion and may soon thereafter commence to senesce.

Whereas the cell number appears to increase exponentially from initiation until unfolding (at least, in a constant environment), the pattern of change of volume of the whole leaf is less certain. The data of Fig. 7.2 indicate that the change is close to exponential until some time after unfolding, with perhaps a slightly slower rate in the earliest stage. Data from other species also indicate an exponential phase but of varying duration; frequently the relative growth rates, especially of late-formed leaves, decline throughout. The dry weight per unit volume probably decreases during growth and is about 0.25–0.3 g cm^{-3} around the time of unfolding. Many of the earliest cells formed in the primordium are destined to differentiate into vascular tissue; as division continues, an increasing proportion of mesophyll cells is formed. The differentiation of provascular strands is evident at an early stage and the differentiation of epidermal, palisade and spongy mesophyll is early discernible. Differential growth rates may occur in differing parts of the leaf and are reflected by the final shape of the leaf; for example, the basic structure of compound leaves can be seen when the leaf is still extremely small. Generally, cell division ceases first in the epidermis – stomata being initiated just before it stops – then in the spongy mesophyll and finally in the palisade. It also appears to stop first in the distal tip, where growth rates are also slower than in the basal regions. In cereals and grasses, cell division and leaf expansion cease first at the tip; moreover, those parts which emerge from the encircling sheaths have already stopped growth so that growth of the whole lamina is complete with the emergence of the ligule. The sheath, however, continues expansion for a longer time.

Although the pattern of leaf growth along a stem axis is often confused by effects of the environment, especially nutrient supply, it seems to be reasonably consistent in both the Gramineae and the Dicotyledoneae. Successive leaves have slower rates of cell division and expansion and hence relative rates of growth (Fig. 7.3). The phase of constant relative rate of cell division is, however, maintained for a longer time, so that at the time of emergence (or unfolding) successive leaves have greater numbers of cells of smaller size. This leads to successive mature leaves with greater numbers of cells, but whether the leaf is larger or smaller depends primarily on cell expansion and, hence, the supply of mineral nutrients to the expanding leaf.

The overall growth (by weight or volume) up to the time of unfolding or emergence can usually be described by a linear function of the logarithms with time, as illustrated in Figs. 7.2 and 7.3. After unfolding, a higher-order polynomial of the logarithms such as

$$\ln \mathrm{W} = a+bt+ct^2+\ldots \qquad 7.1$$

may be adequate; frequently, a good fit is given by the simple logistic

$$\mathrm{W} = A/\{1+b\exp(-kt)\} \qquad 7.2$$

where $A/(1+b) \leqslant \mathrm{W} \leqslant A$ when $0 \leqslant t \leqslant \infty$. For example, the growth of each leaf on the main axis of a cultivar of wheat, grown with adequate light and mineral nutrient supplies at 25 °C, could be approximated by Equation 7.2 where, taking B as the leaf position counting from the base, the two coefficients were given by

$$k = (0.758+0.117\mathrm{B})^{-1}\,\mathrm{d}^{-1}; \quad A = \exp(7.64\text{–}4.4k)\ \mathrm{mg} \qquad 7.3$$

Light and temperature appear to influence cell division in the leaf primordia to much the same degree as they influence leaf initiation, with the notable difference that division in leaf primordia is extremely slow in complete darkness or at very low irradiances. For example, the relative rates of cell division, $(1/\mathrm{N})\mathrm{dN}/\mathrm{d}t$, in the leaf primordia of the plants described in Fig. 7.1*b* are given by $(1/\mathrm{N})\mathrm{dN}/\mathrm{d}t = (0.925+4.24/Q_\mathrm{V})^{-1}\,\mathrm{d}^{-1}$. However, with increasing irradiance, cell elongation is decreased with partial compensatory increases in expansion in the two dimensions normal to the leaf axis. These effects are more pronounced in cereals and grasses than in dicotyledonous species and may reflect differential responses to a phytochrome-mediated reaction. (Generally, leaves of graminaceous plants respond more like stems than do those of dicotyledonous plants.)

Cell expansion also seems to be influenced by the supply of mineral elements, especially of nitrogen. High nitrogen supplies will usually lead to large leaves. The potential for development of a leaf is therefore set primarily by irradiance and temperature but its realization depends on nutrient supply

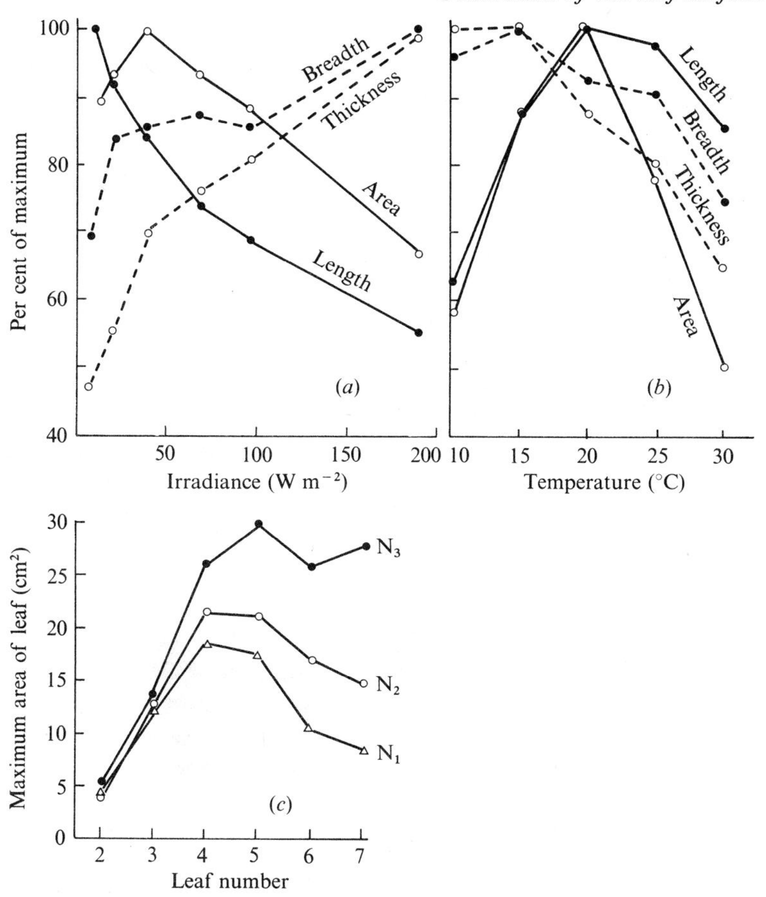

Fig. 7.4. (*a*, *b*) Relationships of length, breadth, thickness and area of wheat leaves with (*a*) visible irradiance and (*b*) temperature. After Friend, D. J. C., Helson, V. A. & Fisher, J. E. (1962). *Can. J. Bot.* **40**, 1299–311. (*c*) Variation in areas of successive leaves of wheat with nitrogen supply on plants grown at high densities in the field where N_1, N_2 and N_3 represent 0, 50 and 235 kg ha^{-1} of applied nitrogen. Data of D. W. Puckridge reproduced in Milthorpe & Ivins (1966).

and direct effects of light and temperature on the expanding leaf. The final weight and volume achieved is likely to increase with increase in irradiance over much of the normal range but the area may reach a maximum at quite low irradiances (Fig. 7.4). All three quantities show optimum curves with respect to temperature, the position of the curve varying with species. For example, the two coefficients for the data described in Equation 7.3 were related to temperature over the range 5–30 °C by

$$A/A_{25} = k/k_{25} = -0.199+0.0974\mathrm{T}-0.001965\mathrm{T}^2 \qquad 7.4$$

The leaves of most species appear to be unaffected by water status over the range of, say, 90–100 per cent relative water content. However, should the relative water content fall below about 90 per cent, cell expansion is reduced, ceasing entirely at 70–75 per cent. Cell division appears to be less affected than expansion in the higher range of water contents but also ceases at much the same values. There is certainly an immediate direct effect on cell expansion due to inadequate turgor pressure to maintain the growth of the cell walls. With prolonged deficits, indirect effects arise from reduced photosynthesis, reduced mineral nutrient supplies and protein synthesis, increased sugar synthesis (at the expense of starch), and other impairments of metabolism.

The changes in physical structure of a leaf during its development are also accompanied by changes in chemical composition and physiological activity. There is, for example, a continued increase in nitrogen and protein, phosphorus compounds, ribose nucleic acid, chlorophyll and cell wall constituents, on both a per leaf and per cell basis, until about the time of maximum area. The total nitrogen and phosphorus contents and their compounds may then begin to decrease although cell wall materials are added until very late in the life of a leaf. There appears to be a comparable pattern in the activity of many enzymes although, of course, the activities of all enzymes do not follow similar paths. These changes are reflected by changes in physiological processes such as photosynthesis and respiration (cf. Fig. 5.18). The rate of photosynthesis per unit area reaches a peak at or some little while before the leaf is fully expanded, whereas the respiration rate declines throughout. In grasses and cereals, the behaviour is more complex: the distal parts of the leaf appear to be less active (on an area basis) than the basal parts and, as all tissue is fully expanded at the time it emerges from enclosing sheaths, the mean rate per leaf reflects the resultant between the ageing of fully expanded cells and the emergence of tissue of inherently greater activity.

Changes in photosynthetic activity could result from changes in any of the components listed in Equation 5.8 (p. 74). In this analysis changes in b might be expected to reflect changes in chlorophyll content or any of the photochemical reactions in photosynthesis. However, b does not vary greatly, most of the changes involving the various resistances in the carbon dioxide pathway. In many dicotyledonous plants, both r'_s and r'_x appear to change, the stomata opening most widely about the time of maximum expansion and thereafter showing successively higher resistances. The carboxylation resistance also appears to increase from about the time of full expansion. In wheat, the major change associated with the decline in photosynthesis appears to be in the carboxylation resistance. In maize, and possibly other C_4 grasses, the rate of photosynthesis remains constant and stomatal and carboxylation resistances stay at near-minimum values for surprisingly long periods after the leaf has fully expanded (cf. also Chapter 5).

The leaf of a dicotyledonous plant depends entirely on the supply of carbohydrates from older leaves until after unfolding. It probably does not become completely self-sufficient until it is 0.2–0.3 its full size and, even thereafter, it may continue to import carbohydrate although it is a net exporter. Its peak period of transport is around the time it reaches full size (cf. Fig. 5.18). Net import of nitrogen, as NO_3^- or amino acids, phosphorus and potassium continues until the leaf reaches its full size. In the earliest stages most of these seem to come from older leaves, but as the leaf grows higher proportions are imported directly from the roots. It often becomes a net exporter of these major nutrients around the time of full expansion; the decrease in physiological activity is no doubt associated with these losses and denotes the beginning of senescence. The wheat lamina imports all of its carbohydrate until emergence and even at full emergence is importing about one-quarter of its dry weight increase from lower leaves (especially the second leaf below). About half of the dry weight of the sheath comes from sources other than the attached lamina.

In later stages of senescence – i.e. *progressive senescence* as it may be called to distinguish it from the *flowering senescence* which accompanies seed maturation in monocarpic stem axes – there is also marked loss of chlorophyll, gradual disruption of organelles and all membranes, loss of water, and finally the browning and death of the leaf. Senescence is usually hastened by conditions which favour rapid growth and often appears to involve a demand for the major mobile elements which cannot be met by absorption through the roots.

THE GROWTH OF STEMS

The primary growth of the stem arises from cells initially produced by the apical meristem. In some species cell division in the sub-apical meristems continues for a short time (possibly only 4–5 plastochrons) but in others dividing cells may be found as far as 10 cm below the apex. It usually proceeds at a slower rate than in the leaf primordia. The cells at this early stage are about 5–6 times the volume of those of the leaf primordia, have a larger content – but a lower concentration – of protein, and are more active metabolically. It has been suggested that they may synthesize some of the metabolites required by the primordia. Extension of these cells is at first radial only, there being very little growth in length during the first few plastochrons. During this time there is differentiation first of the procambium, then of the phloem and finally of the xylem, these progressing acropetally from the older bundles of the stem. Later, the cells elongate extensively along the main axis, leading to growth of the internodes and increased

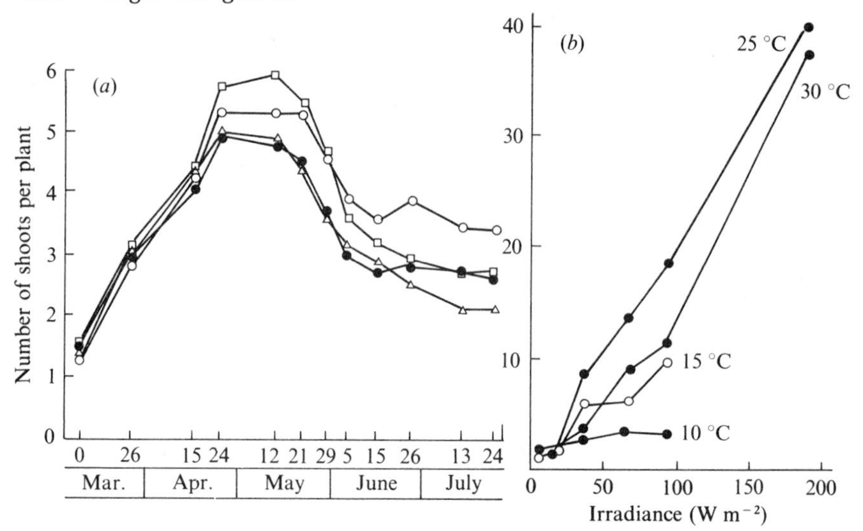

Fig. 7.5. (*a*) The course of tiller numbers in four varieties of wheat grown in southern England. Data of S. Kirinde reproduced in Milthorpe & Ivins (1966). (*b*) Tiller production as a function of light intensity and temperature of 4-week-old wheat plants. Data of D. J. C. Friend reproduced in Milthorpe & Ivins (1966).

height of the stem; secondary growth may follow as a result of cambial activity.

The rates of stem elongation vary widely between species, maxima of about 25 mm h^{-1} being recorded (in bamboo). A rate 10^{-1} to 10^{-2} times this value would be more usual. Elongation rates show the usual 'optimum' response curve to temperature and are inversely related to irradiance; the latter appears to be a phytochrome-mediated response and to vary widely between species. Because of this response, plant height is a very poor, although widely used, measure of growth in general. In most cereals and grasses, there is virtually no elongation of the stem until after initiation of the inflorescence; the upper 4–5 internodes then grow extensively and at the time of anthesis the stem may comprise as much as one-quarter to one-third the total dry weight of the shoot (cf. p. 143). (Stem elongation is not causally associated with inflorescence growth, however, because it can be stimulated in some grasses by suitable photoperiods in the absence of flower induction.)

The ability of the plant to produce branches (i.e. of the axillary buds to recommence growth) varies from none, or at most two, in sunflower and maize, to the profusion which is the characteristic feature of perennial pasture grasses. Branching behaviour may be illustrated by reference to the small-grain cereals such as wheat and barley. The characteristic pattern in a field crop is, first, the production of a number of branches (tillers), and

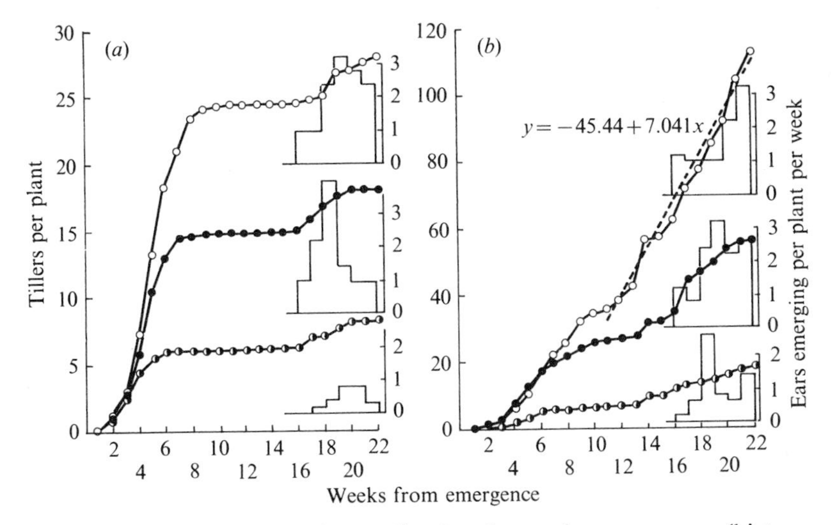

Fig. 7.6. Tiller production (continuous lines) and rate of ear emergence (histograms) of barley plants grown in solution culture with (*a*) all nutrients supplied before germination and (*b*) nutrients replaced weekly. In (*a*) and (*b*) respectively, open circles represent solutions of 1 and 0.5, closed circles of 0.5 and 0.05, and pied circles of 0.1 and 0.01 times a standard concentration of all nutrients. After Aspinall, D. (1961). *Aust. J. biol. Sci.* **14**, 493–505.

this is followed by the death of an appreciable proportion of these without forming inflorescences (Fig. 7.5*a*). The rate of production and maximum number of tillers produced vary greatly between cultivars and are markedly influenced by the supply of photosynthate (Fig. 7.5*b*) and mineral nutrients (Fig. 7.6). Although much evidence indicates that auxin and an unidentified inhibitor, at least, participate in the internal control reactions – variations in these may possibly reflect the large differences between species and cultivars – the variation within a cultivar is dominated by supply of substrates, especially nitrogen. The mechanism concerned with the later survival – or lack of survival – of tillers is uncertain. The bud depends on assimilate from the older leaves during the early stages of its resumed growth but, with the emergence of its own leaves, becomes independent and there is then very little import, at least of carbon compounds, from other parts of the plant. Death appears to occur because of an insufficient supply of substrates to the apex and primordial leaves, and these are the organs which die first. However, if the tiller survives until the apex has been induced to flower, death does not then occur and the normal sequence of inflorescence differentiation and grain-filling proceeds. Successive tillers are smaller and yield successively smaller amounts of grain, possibly because they develop with lower mineral and assimilate supplies. In addition, the late-formed

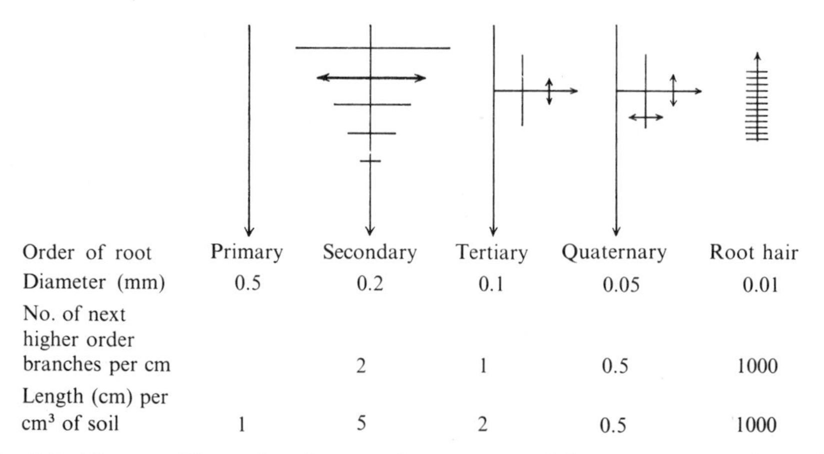

Order of root	Primary	Secondary	Tertiary	Quaternary	Root hair
Diameter (mm)	0.5	0.2	0.1	0.05	0.01
No. of next higher order branches per cm		2	1	0.5	1000
Length (cm) per cm^3 of soil	1	5	2	0.5	1000

Fig. 7.7. Diagram illustrating the constituent parts of the root system of a well-established cereal crop with dimensions relative to the top layers. Adapted from Barley, K. P. (1970). *Adv. Agron.* **22**, 159–201.

tillers are induced to flower early in their ontogeny and their inflorescences develop quickly.

ROOT GROWTH

The primary root and, in monocotyledonous plants, up to four further seminal roots, are differentiated in the embryo. During germination these grow as a consequence of cell division in their apical meristems and the subsequent extension of these cells. Cell division in the root is usually more rapid than in the leaf primordia, which in turn is greater than in the stem. The number of dividing cells in any one root meristem appears to be maintained for a long time, so that under the same conditions the rate of elongation of a root axis remains constant with time. The zone of cell elongation may extend to 0.5–1.5 cm from the tip; at the older (proximal) end root hairs arise – under favourable conditions from alternate epidermal cells but often much less frequently – and attain a length of 0.5–1.5 mm. Excluding about the first centimetre of each primary axis and its branches, laterals of the next higher order arise, finally producing a much-branched and ramifying root system (Fig. 7.7). The primary roots usually grow mainly vertically downwards, while secondary roots may grow in a more horizontal direction for several centimetres before turning downwards, and higher-order laterals follow a more or less random path.

The main roots elongate at rates of, say, 1–20 mm d^{-1}, eventually reaching a length of 1–2 m in herbaceous annuals. Higher-order laterals elongate at slower rates; some evidence suggests that the rates of elongation of primary, secondary, tertiary, ... roots may be approximated by the series 1, $\frac{1}{4}$, $\frac{1}{8}$,

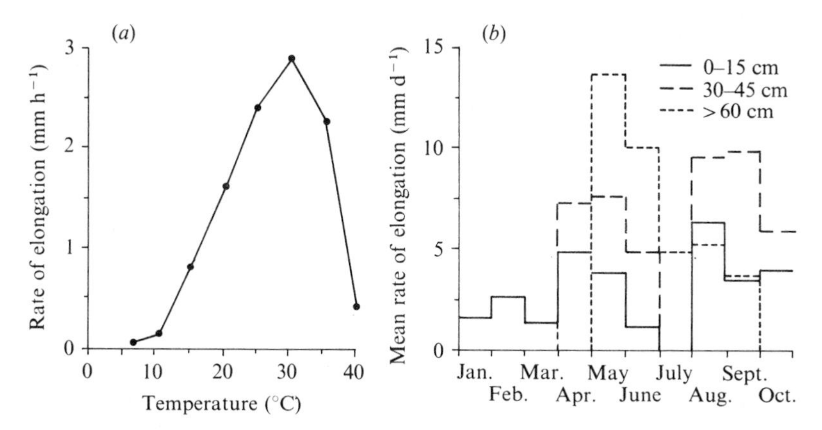

Fig. 7.8. (*a*) Relationship between the rate of elongation of seminal roots of maize and temperature. After Erikson, R. O. (1959). *Am. Nat.* **93**, 225–35. (*b*) Mean rate of elongation of roots (all orders of branching) in a perennial rye-grass sward in the United Kingdom in relation to season and depth. Redrawn from data of T. E. Williams given in Whittington (1969). Note differences in scales.

$\frac{1}{16}$, ... The rate of elongation is, however, related to temperature by the usual optimum relationship (Fig. 7.8*a*). It is retarded by water shortage and markedly influenced by photosynthate supply (and possibly growth substances) from the leaves, severe defoliation leading to virtual cessation of elongation. The effect of nutrient concentration in the soil solution is much less clear. Some evidence suggests that the rate of elongation during seedling stages is retarded by nutrient concentrations which are sub-optimal for shoot growth. Possibly the most general effects arise indirectly from leaf growth and supply of photosynthate to the roots. Branches arise at more or less constant intervals along a root (Fig. 7.7), the length of these intervals varying between species and order of the branch but being little influenced by the environment. However, there appears to be a contradiction between the finding that mineral concentration in the soil solution has little effect, and the frequent observation that root development is most profuse in the soil layers where fertilizer has been applied.

A general paucity of experimental investigations on root growth is offset by a number of laborious observations on root systems made by a few indefatigable workers. This leads to a situation in which, despite lack of understanding of mechanisms and the resulting confidence that stems therefrom, the extent and disposition of root systems should be predictable with reasonable accuracy. In a deep, evenly textured soil, for example, it should be possible to forecast the development of the root system of, say, a cereal crop, given values for the parameters mentioned above. With cereals a major determinant is the rate of differentiation of adventitious roots from the nodes, and very few observations have been made on this. It would seem that the first nodal roots are initiated soon after emergence of the seedling, that the rate of initiation is closely correlated with tiller development (about 5–6 being formed per tiller irrespective of the number of tillers), and that new roots continue to be formed at least until anthesis. Some evidence suggests that, with a profuse development of adventitious roots, the seminal root system may senesce. There is certainly continued initiation and death of roots in swards of perennial grasses, where the average life of any one root varies with the time of year when it was initiated, usually being about six months. In the field, the initiation of new roots has been found to be negatively correlated with soil temperature, i.e. an opposite effect to that on the rate of elongation of existing roots. There are also large variations with season, time of cutting of sward, and soil depth in the rate of elongation (Fig. 7.8*b*). Seasonal variations are also found with other perennial crops. Roots of fruit trees usually show two periods of intense activity, one in spring and the other in late summer; root growth always commences in the spring before shoot growth is evident. In such species, of course, the primary and secondary roots persist for many years; they exhibit secondary thickening and form the framework of the root system, usually permeating the soil to 1½–2 times the projected limits of branch spread.

In a deep, open-textured soil, root systems develop as outlined above, reaching concentrations of up to 50 cm^{-2}. Rarely is this ideal achieved (cf. Table 3.2). Differences in compaction, often accentuated by tillage implements, surface cultivation to control weed growth, the activity of nematodes, and other factors, lead to uneven distribution and less intense development than is potentially possible. Whether or not this is significant in determining the yield of the product for which the crop is grown depends on the resultant of a complexity of components, including the actual concentration and distribution of roots in the various soil layers, the availability of mineral nutrients therein and the frequency of rewetting, and can only be assessed by the development of the type of approach outlined in Chapter 10.

GROWTH OF SWOLLEN STEMS AND ROOTS

In a number of species particular regions of stems or roots show exaggerated radial growth, these forming the harvested products of the crop. Here, we will mention only two – the potato, grown for its swollen stems or tubers, and those forms of *Beta vulgaris* (sugar beet, table beet and mangold) grown for their swollen tap roots.

Potato tubers

Tuber formation is first detectable when the cells of the youngest expanding internode on the underground diageotropic stems (stolons) commence to expand radially rather than extend principally along the stolon axis. At this stage the terminal bud consists of about twelve leaf primordia with buds in the axes of most. Very soon cell division commences in all tissues of the young tuber, which continues to grow by the production of new leaf primordia, buds and internodes from the apical meristem, and the division and expansion of cells in the expanding internodes, the differentiation of new leaves from the apical meristem keeping pace with internode expansion. This activity is finally arrested by the death or destruction of the above-ground shoots and at this time the terminal and lateral buds normally enter a phase of dormancy. Leaf growth is, of course, arrested on stolons as well as tubers at a very early stage; the axillary buds continue to grow for some time, producing new leaves and buds, but these also become dormant. Most of the tuber volume arises from the secondary cell divisions proceeding throughout all of the tissues. In one study, tubers with a volume of 2.5×10^{-2} cm^3 at initiation contained 2.68×10^5 cells; when the volume was 300 cm^3 they contained 366×10^5 cells. That is, the 12000-fold increase in volume was accompanied by a 137-fold increase in cell number and an 88-fold increase in mean cell size (see also Fig. 9.7).

Numerous stolons and stolon branches are produced, and these initiate tubers over a long period of time. Tuber initiation is favoured by short days, shortage of nitrogen and phosphorus, high radiation and low temperatures; the various internal factors which lead to initiation, however, remain unknown. Even though tubers continue to be initiated, only those formed during the first two weeks or so grow to marketable size. The individual tubers grow at different rates at different times, indicating some complex form of control. One of the factors would seem to be photosynthate supply, as the mean size of tubers is an inverse function of the number of tubers initiated and a positive function of the amount of foliage developed. Tuber growth ceases only with the death of all the foliage. Despite the variations between individuals, the population of tubers as a whole grows at a rate which is remarkably constant with time and independent of variations, over a reasonably wide range, in radiation, temperature and area of leaves (cf. Fig. 9.11). That is, the rate of increase in tuber weight is determined at the time of tuber set and its duration depends on the maintenance of photosynthetically active leaves. Most of the tuber weight comes from current photosynthate, substances elaborated before tuber initiation and stored in the stem contributing about 10 per cent of the whole.

This progression can only be upset by environmental variations of a catastrophic proportion, such as frost, disease, or an appreciable water deficit. The tubers, as do developing succulent fruits in other species, provide a partial buffer in the water supply to leaves; the transfer of water to the leaves often leading to a decrease in volume during the day but with a subsequent increase again during the night when transpiration ceases. However, tuber expansion may be completely interrupted by a severe deficit;

under such circumstances the internodes may continue to elongate but not expand radially. When the water supply is restored and the tuber resumes radial expansion, it is only the new internodes which participate; the previously radially expanding tissue and the longitudinally elongating internodes do not grow.

Beet roots

The storage roots arise from the formation of some 8–12 concentric cambia in place of the usual single cambium involved in secondary growth. All of the cambia are initiated soon after germination and growth then proceeds by continued cell division in these regions. The very large increase in volume over the growing season is due entirely to new cells, the average cell volume staying remarkably constant at about 10^{-7} cm^3 (in sugar beet). As there are about 10^9 cells per g dry weight of storage root, a freshly harvested root weighing 0.5 kg may contain more than 10^{11} cells.

Growth of storage roots, absorption roots and leaves progresses concurrently throughout ontogeny, with the storage root obtaining an increasing proportion of assimilates with age. (Stem growth is suppressed.) Partitioning of growth varies between the forms, those cultivars selected for sugar production (sugar beet) usually having a lower ratio of root to leaves and a higher proportion of stored sugar than those selected for stock feed (mangold); the latter appear to have been selected to produce the maximum weight of root possible, irrespective of composition. With any one cultivar, the pattern of growth remains much more stable than is found in cultivars of other species; in particular, the senescence of the older leaves is much more gradual (but see Figs. 9.8 and 9.9).

FURTHER READING

Eastin, J. D. *et al.* (eds.) (1969). *Physiological aspects of crop yield.* American Society of Agronomy, Madison, Wisconsin.

Evans, L. T. (ed.) (1963). *Environmental control of plant growth.* Academic Press, New York and London.

Humphries, E. C. & Wheeler, A. W. (1963). The physiology of leaf growth. *Ann. Rev. Pl. Physiol.* **14**, 385–409.

Ivins, J. D. & Milthorpe, F. L. (eds.) (1963). *The growth of the potato.* Proc. 10th Easter Sch. agric. Sci., Univ. Nottingham. Butterworths, London.

Langer, R. H. M. (1972). *How grasses grow.* Edward Arnold, London.

Leopold, A. C. (1964). *Plant growth and development.* McGraw-Hill, New York.

Milthorpe, F. L. (ed.) (1956). *The growth of leaves.* Proc. 3rd Easter Sch. agric. Sci., Univ. Nottingham. Butterworths, London.

Milthorpe, F. L. & Ivins, J. D. (eds.) (1966). *The growth of cereals and grasses*. Proc. 12th Easter Sch. agric. Sci., Univ. Nottingham. Butterworths, London.

Sachs, R. M. (1965). Stem elongation. *Ann. Rev. Pl. Physiol.* **16**, 73–96.

Steward, F. C. (1968). *Growth and organization in plants.* Addison-Wesley, Reading, Mass.

Steward, F. C. (ed.) (1969). *Plant physiology*, vol. 5A. Academic Press, New York and London.

Whittington, W. J. (ed.) (1969). *Root growth*, Proc. 15th Easter Sch. agric. Sci., Univ. Nottingham. Butterworths, London.

Williams, R. F. (1955). Redistribution of mineral elements during development. *Ann. Rev. Pl. Physiol.* **6**, 25–42.

Woolhouse, H. W. (ed.) (1967). *Aspects of the biology of ageing. Symp. Soc. exp. Biol.* **21**. Cambridge University Press.

8

Flowering and Fruit Growth

FLOWERING AND INFLORESCENCE GROWTH

The change of an apex, whether apical or axillary, from the vegetative to the flowering state represents the outcome of an internal autonomous progression which is modulated to varying degrees, depending on cultivar, by low temperature and photoperiod. The progressive tendency towards proneness to flower with increasing age is shown in some species by flower production only after several seasonal cycles, and in others by the requirement for a decreasing number of cycles of exposure to an inductive photoperiod with increase in size.

The change in state of the apex is a profound one, resulting in the production of completely dissimilar structures and almost certainly represents the repression of one group and the evocation of another group of genes. Evidence given in most texts of plant physiology shows that some species, such as *Beta vulgaris* and *Brassica oleracea*, or groups of cultivars of a species, such as winter wheat, have a requirement for a period of low temperature (2–4 °C) before they will flower; this vernalization response is perceived directly by the apex. (Other species, such as most deciduous fruits, require a period of low temperature in order to break the dormancy of flower buds already differentiated.)

Other cultivars respond to photoperiod, this being perceived by the leaves and a message (of unknown nature) transmitted to the apex where flowers are initiated. The usual classification is into strict long-day plants requiring a night length of less than 12 hours, or strict short-day plants, requiring a night length of 8 hours or more before they will flower. Between these and the day-neutral varieties, which have no requirement, is a large spectrum of facultative long- and short-day strains, flowering being hastened the shorter or longer the nights, respectively. It is to these last two groups that most cultivars belong but this is not to say that photoperiod is not important. Some rice varieties, for example, are sensitive to photoperiodic differences of the order of 20 minutes and the suitability of soybean varieties for a region is closely related to the photoperiod.

From the nature of these responses, there follows the general principle that the more favourable the photoperiod or vernalization conditions then

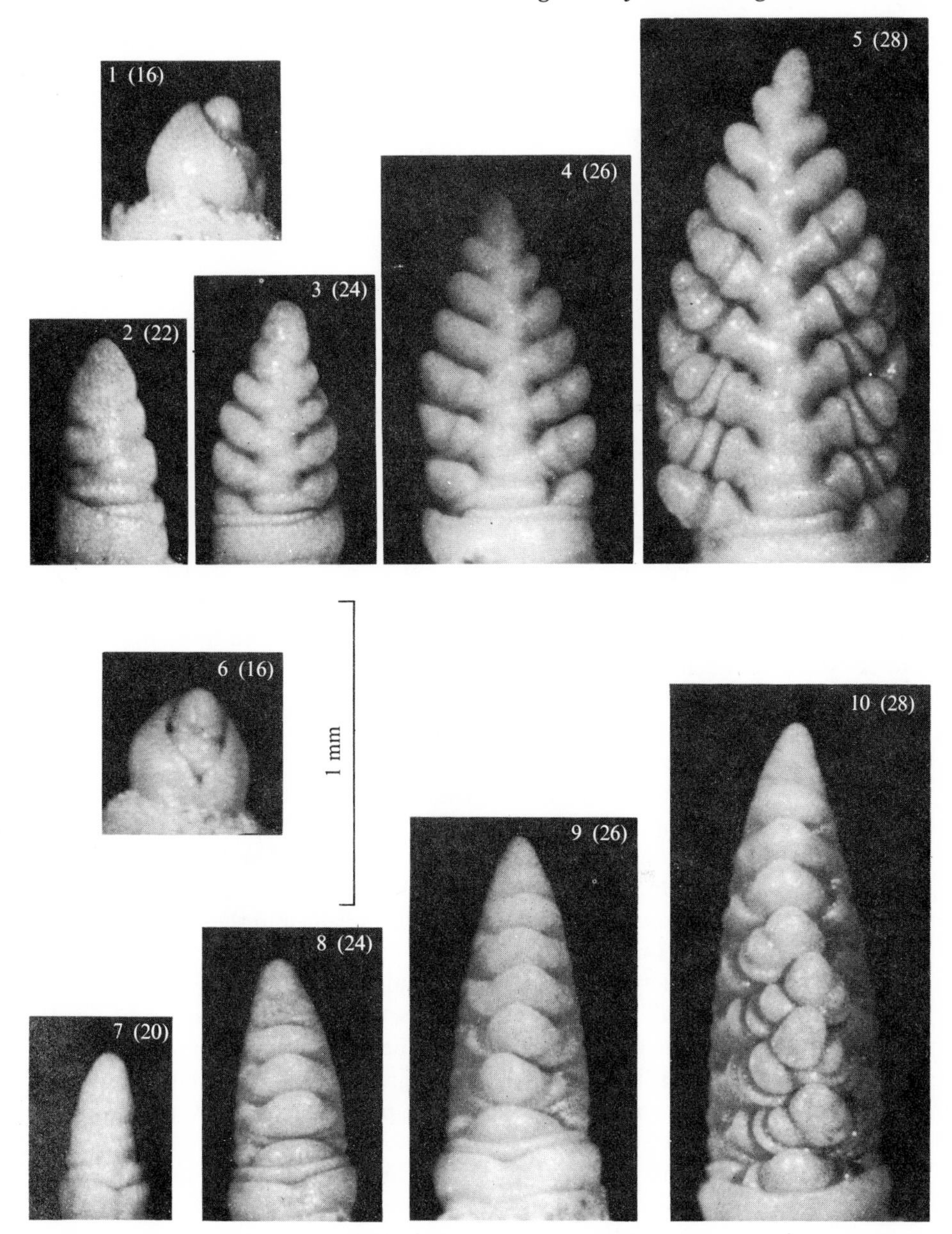

Fig. 8.1. Early stages in the development of the inflorescence of wheat. The figures in parentheses state the number of days from sowing. The photographs are taken from two angles, one normal to the other. 1, 6: vegetative apex; 7: prior to double ridge formation; 2: double ridge formation; 3, 8: spikelet primordia developing; 4, 9: glume ridges developing on more advanced spikelets; 5, 10: flower primordia developing in the more advanced spikelets. After Williams, R. F. (1966). *Aust. J. biol. Sci.* **19**, 949–66.

the earlier and fewer are the flowers formed; conversely, the less favourable the conditions then the longer it is before flowers are initiated and the greater the number of flowers formed, especially in terminal inflorescences. That is, the flowering requirements of early varieties coincide with the particular photoperiod obtaining, whereas those of late-season varieties do not. With tuber-forming species such as the potato, the high-yielding late-season varieties are those with a substantial short-day requirement grown in a season of long days.

In terms of histogenesis, the transition of the apex from leaf to flower production is marked by a change from tunica-dominated to corpus-dominated growth. There is then a surge in activity of the inner cell layers relative to the outer layers, leading to the transformation of the apical cone from a rather flat to an elongated acute shape in wheat (potential spike) or a large sphere in *Chrysanthemum* (potential capitulum). It appears that the potential bracts, carpel and ovule integuments are derived, as are leaves, from the outer tunica layers, whilst spikelets, florets, stamens and ovules arise from all tissues, as do axillary buds (cf. p. 124).

The flowering structures vary widely between species, of course, from single axillary flowers as in peas to complex terminal and limited inflorecences as in wheat and sunflower, indefinite terminal racemes as in most brassicas, to consecutive indefinite terminal racemes in tomato (the uppermost axillary bud growing out after the flowering evocation of the terminal meristem). We will follow, very briefly, only one example – the wheat spike (Fig. 8.1). In the vegetative state (Fig. 8.1, 1, 6), the hemispherical apical dome can be seen, with the ultimate leaf just visible and the penultimate leaf growing around the entire diameter of the stem. Photographs 2 and 7 show the formation of double ridges, the earliest detectable signs of spikelet primordia. These are more clearly shown in photographs 3 and 8. As time progresses, there is continued formation of new spikelet primordia and these differentiate to form glume ridges (4, 9) and, a little later (5, 10), flower primordia in the most advanced spikelets – here, the sixth and seventh from the basal end. Shortly afterwards, stamen differentiation is visible in the most advanced spikelet and, after this time, there is no further formation of new spikelets. Each of the existing spikelets continues to grow and differentiate, forming up to 5–7 florets; successive florets grow at progressively slower rates and reach smaller ultimate sizes.

Under constant conditions, the dry weight and volume of the inflorescence increases exponentially until just before emergence of the ear from the enclosing leaf sheaths; both then decrease sharply. Growth in length usually shows two phases: an initial period of slow elongation during the time of formation of new spikelets (i.e. during the stages depicted in Fig. 8.1) followed by a period of rapid elongation of the internodes of the rachis. The latter appears to be associated with increased production of gibberellins.

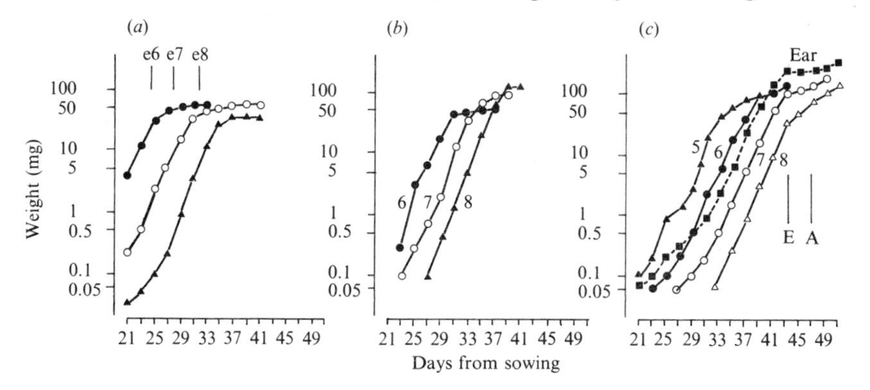

Fig. 8.2. Relative rates of growth of (*a*) laminae, (*b*) leaf sheaths and (*c*) internodes and ear during the period of inflorescence development of wheat. The numbers indicate the sequence of development, e the time of emergence of the particular lamina, E emergence of the ear and A anthesis. Note logarithmic scales. After Patrick, J. W., Ph.D. thesis, Macquarie University.

Stem growth (cf. p. 131) commences from the time of inflorescence initiation. It is convenient to regard the cereal shoot as being made up of a series of units, each of which consists of a node, axillary bud, an internode above and a leaf subtended from its node. That is, the leaf sheath of unit *n* surrounds internode *n* for part of its length when mature. In any one unit, development proceeds in the order: lamina before sheath before internode (Fig. 8.2). Only the upper few internodes elongate, the early development of the inflorescence being more or less coincident in time with that of the antepenultimate internode, and its most active growth synchronous with the growth of the last two internodes. Growth of the upper two internodes continues until 1–2 weeks after anthesis; grain growth starts soon after anthesis but the other parts of the inflorescence also continue growing for some further 2–3 weeks. Because the upper three internodes develop over the period that the inflorescence is growing, they provide potential competitive sinks for assimilate and mineral nutrient supplies. However, information from the flow of ^{14}C-labelled compounds shows that as much moves into the inflorescence from any source lamina as into the two internodes immediately below (Fig. 8.3). Each succeeding leaf contributes progressively more to ear growth, almost half of the pre-anthesis weight of the ear coming from the flag leaf. (As indicated later, the flag leaf is also by far the most important source of carbohydrate during grain filling.)

The potential yield is determined during the early stage of inflorescence differentiation by the number of tillers which form inflorescences and the number of spikelets (and florets) formed per inflorescence. Although there is a dearth of relevant information, it seems that the environment influences these components through two opposing effects – one working through the

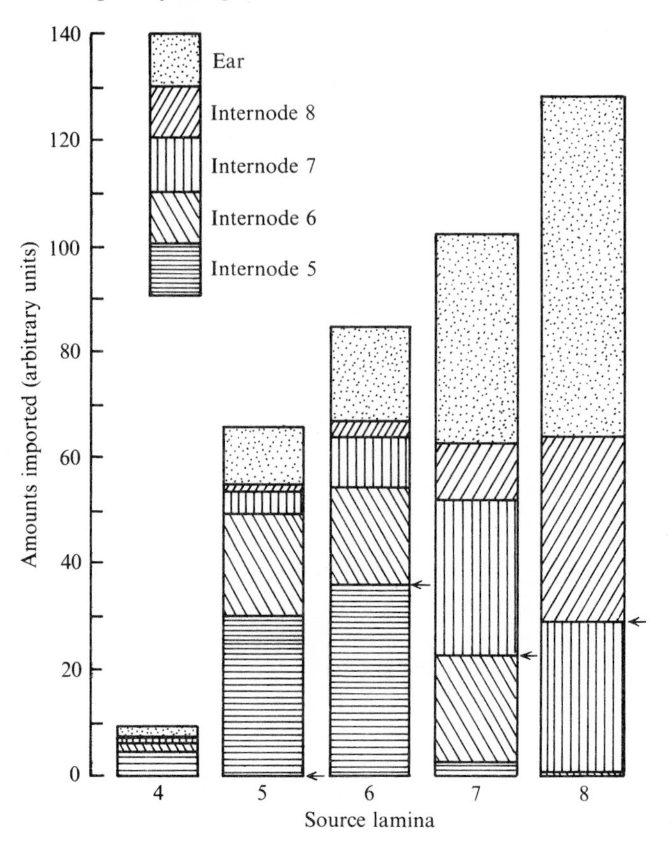

Fig. 8.3. Relative amounts of carbohydrate imported by the various sinks from the different source laminae over the period of inflorescence development of wheat from initiation to anthesis. These data come from the same experiment and cover the same time period as those in Fig. 8.2. The arrows indicate point of attachment of the source lamina concerned. After Patrick, J. W. (1972). *Aust. J. biol. Sci.* **25**, 455–67.

growth and development of each spikelet and the other on the number of spikelets formed. Spikelet differentiation appears to stop, because of some unknown internal control mechanism, when stamens are differentiated in the most advanced floret. It follows, therefore, that the greatest number of spikelets will be formed when the conditions for growth of the spikelet are least favourable. The relative rates of spikelet formation and spikelet development are such that low temperatures result in long inflorescences with more spikelets, but a longer time is required, than at optimal temperatures. Increased light intensity, or rather rate of supply of photosynthate to the developing inflorescence, appears to affect both rates more or less equally but, in addition, the number of flowering tillers is increased at higher intensities. The magnitude of these effects, however, varies markedly between cultivars depending on their vernalization and photoperiod requirements. Influences attributable to the supply of mineral nutrients are even less clear; by analogy with the effects on leaf growth, it seems possible that differences in supply have little effect early in the growth of the inflorescence but can modify the size reached by each floret. More far-reaching influences are likely to be exerted through the number of fertile tillers produced and, indirectly, through the extent of the leaf surface generated.

The effect of a water deficit is less certain. During the early stages of inflorescence development, when cell division is paramount, a transient water shortage appears to suspend rather than to completely disrupt differentiation. As in all tissues, cell division appears to be less influenced by a given water deficit than is cell extension; hence, although the rate of formation of primordia may be reduced, it recovers after restoration of the water supply. If the period of water deficit is short, therefore, and at an early stage of inflorescence development, there may be a delay in spikelet formation but no reduction in size; if it continues or occurs late in the period of development, then both the number and size of spikelets may be greatly reduced. Cereal inflorescences appear to be particularly vulnerable at a late stage of development – between ear emergence and anthesis – when death of many spikelets may result. They are also susceptible at anthesis, when germination of pollen and growth of the pollen tube are greatly affected.

The rate of growth of the inflorescence and its component florets decreases as it reaches anthesis. By about this time, the subtending bracts, carpels and stamens have reached full size, although the ovule and possibly the ovary may continue to grow at a slow rate until the resurgence of activity which follows fertilization. The peduncle and internodes of the rachis also continue to elongate at least until towards the end of anthesis. It might be expected, at least in a strictly determinate plant, that growth of the whole would fall to a low level between the end of flower growth and the beginning of seed and fruit growth. Although this has sometimes been noted, it is difficult to detect because of differences in the times of development of flowers within an inflorescence, of inflorescences on a plant and between plants in the different samples that need to be taken.

Inflorescences of few species have been described as thoroughly as that of wheat, but even with wheat there is little information on quantitative interrelationships. Large differences in some of the controlling mechanisms are indicated by the wide range of inflorescence types found. In cotton, for example, there appears to be no effect of the most advanced flower in checking further flower production; these continue to be produced even after some of the fruits have matured. In these indeterminate or indefinite-flowering species, abscission of flowers already produced rather than cessation of initiation appears to be the main way of controlling the number.

SEED AND FRUIT GROWTH

Growth of the ovary may continue at a perceptible rate up to and beyond anthesis, or it may cease transiently and resume growth following germination of the pollen. When the pollen tube reaches the ovule and fertilization occurs, the rate of growth increases markedly. In many species, if fertilization does not occur, the ovary then gradually shrinks and the flower may abscind. However, in parthenocarpic species, the fruits develop without fertilization and seed growth. These usually have a large number of ovules and include species in which fruits develop without pollination, such as banana and pineapple, those in which development is initiated by growth of the pollen tube which fails to reach the ovule, such as fruits of some triploid plants, and those in which the seeds abort before reaching maturity, such as some cultivars of cherries and grapes.

Growth of the embryo(s), which with the ovule wall forms the seed, and of the ovary, which forms the outer integuments of the fruit, proceed more or less concurrently (Fig. 8.4). However, integument growth usually ceases before seed growth – sometimes long before, as in the example of pea cited. Both increase markedly in dry weight, water and volume, but as time progresses there is a decrease in dry weight of the hull, representing some transfer to the seeds and a negative balance between respiration and hull photosynthesis plus import. The water content of the hull and later that of the seeds also decreases; this actual loss of water which occurs in all seeds (but not all fruits) is associated with the replacement of osmotic materials by starch and other large molecules with a low absorptive capacity for water. This is not apparent in the data given in this example, which covered the period only to late 'commercial' maturity – not to the full biological maturity of the fruit. Changes in the major constituents are also shown (Figs. 8.4*c*, *d*). It will be noted that at Day 40 only about three-quarters of the dry weight of the seed was accounted for by the constituents measured. The bulk of the remainder consists of cellulose and other cell wall materials. The period of rapid starch development and decrease in sugars was associated with a marked increase in starch phosphorylase.

There are usually large differences between sizes of ovaries at different positions and in the rates at which these (and also ovules within a fruit) develop. This results in an appreciable variation in fruit and seed size at maturity; nevertheless, the *average* seed size of any one cultivar stays remarkably constant over a wide range of environments compared with seed number. That is, genetic factors predominate in determining seed size, a water deficit being the only environmental component to exert much effect. The sizes of fruits, especially multi-seeded or fleshy fruits, however, are much more subject to environmental influence than is seed size. Most environmental effects during fruit growth are exerted on the numbers of

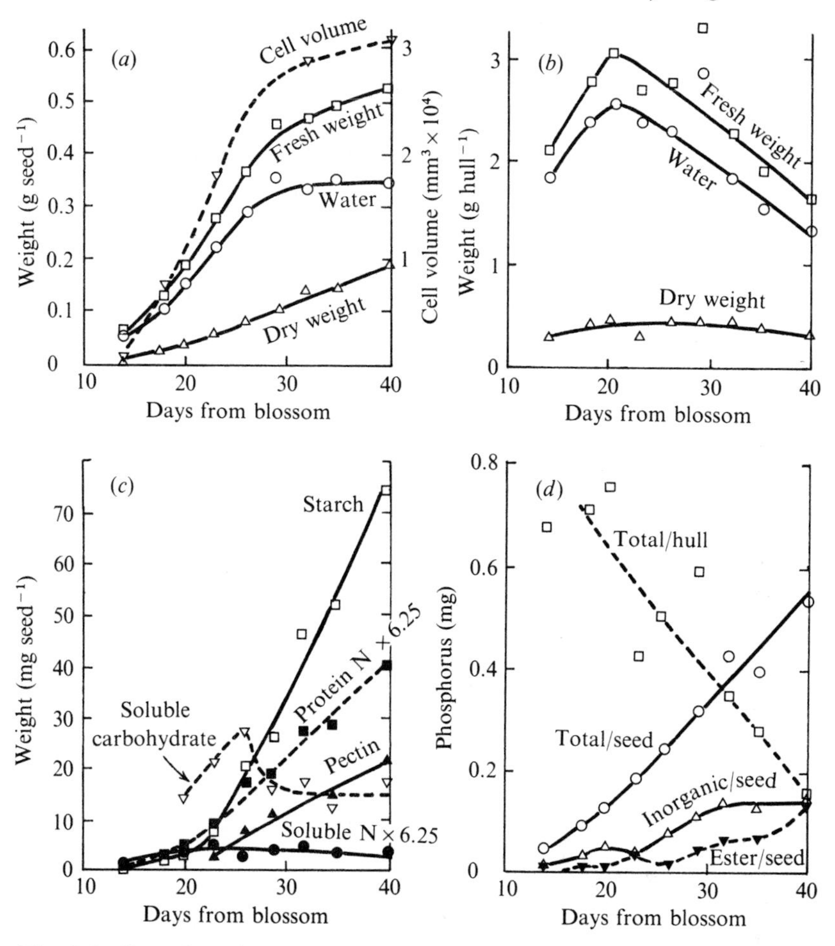

Fig. 8.4. Growth and development of the fruit of pea (cv. Canner's Perfection) from soon after fertilization to just beyond the stage suitable for canning. (*a*) Changes in the seed. Note there are about 6 seeds per pod. (*b*) Changes in the hull. (*c*) Changes in the carbohydrate and protein fractions of the seed. (*d*) Changes in the phosphorus components of the seed. After McKee, H. S., Robertson, R. N. & Lee, J. B. (1955). *Aust. J. biol. Sci.* **8**, 137–63.

fruits or seeds finally produced. Nevertheless, there are differences in response which are related to the variation in time at which all flowers on the plant reach anthesis. In a strictly determinate plant, such as sunflower, most of the flowers develop over a short period of time and set seed, all of which grow at a rate which depends on the environmental conditions (Fig. 8.5). At the other extreme, where flowering extends over a long period, as in mustard, tomato or many cultivars of pea, seeds develop to much the same

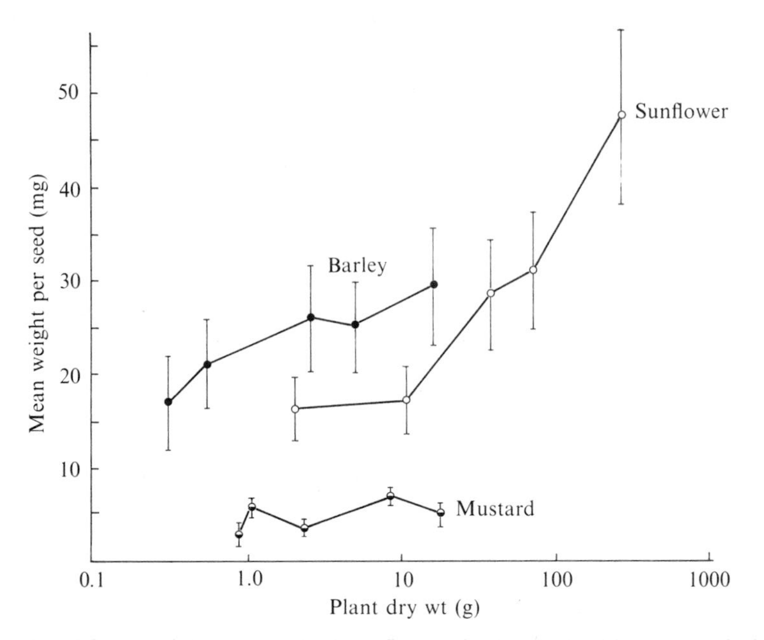

Fig. 8.5. Mean weight per seed of sunflower, barley and white mustard showing variation with carbohydrate and mineral nutrient supplies as induced by differences in spacing; these are reflected by the dry weights in the plant. The vertical lines show the standard deviations of the respective means. Unpublished data of Dr H. Idris.

size despite a large variation in photosynthate and nutrient supply. If growing conditions are poor, the later-formed flowers or fruits simply fail to develop, abscission of flowers and fruits occurring over a reasonably wide range of stages of development. With intermediate species, such as barley and wheat, there are effects on fruit set and on fruit development. Almost invariably, the grains from the basal and proximal spikelets are smaller than those from the large central group (Fig. 8.6). The first and second florets, in a spikelet with a total of about five, usually produce the largest grains and the distal florets usually do not form grains. Denoting the florets in a spikelet from the base as *a–e*, the ovary size at anthesis is in the order $a > b > c > d > e$. At maturity, under optimal growth conditions, grain size may be in order $c > b > d > a > e$. Under less suitable conditions, it is likely to be $b > a > c$ and under poor conditions $a > b$, the remaining florets aborting. How these effects are brought about is not clear; almost certainly some unknown hormonal mechanism is involved and there is also evidence that assimilate and nutrient supplies to the growing parts are implicated.

The carbon compounds that move into the growing fruits and seeds may be formed directly from photosynthesis in the fruit itself or its associated

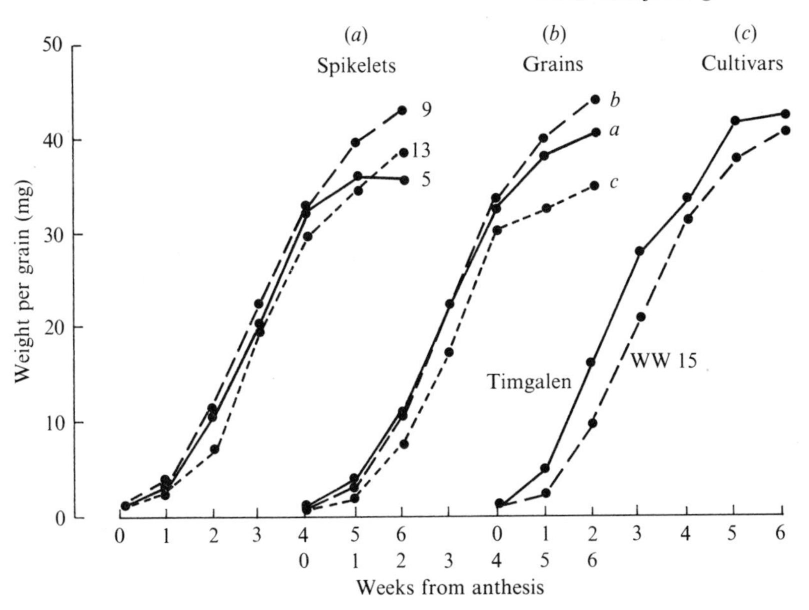

Fig. 8.6. Growth of the grains of wheat as influenced by (*a*) position of spikelet (averages of grains *a* and *b* in spikelets numbered from the apex to the base); (*b*) position of floret lettered *a*–*c* from base to apex of spikelets 7, 9 and 11; and (*c*) cultivar (based on means of grains *a* and *b* in spikelets 9 and 11). After Bremner, P. M. (1972). *Aust. J. biol. Sci.* **25**, 657–68.

appendages (such as glumes and awns of a cereal inflorescence), from photosynthesis in the leaves still extant, or from previously elaborated photosynthate stored elsewhere in the plant. The relative proportions contributed by these different sources vary between species. Almost all of the dry weight of the ear of barley, for example, comes from current photosynthesis, mostly in the ears and flag leaf (Table 8.1). In other cereals, especially awnless varieties, a higher proportion is contributed by the flag leaf, photosynthesis in the ears of many wheat varieties, for example, not even compensating for the amount used in respiration. There is some evidence that in some rice varieties the carbohydrate previously stored in the stem contributes a significant proportion. In addition, the maintenance in rice of two or more active leaves until grain ripening indicates that they may also make an important contribution. With maize, because of its morphology and maintenance of leaves, almost all the carbohydrate of the ear comes from current photosynthate, about half being elaborated in the top six leaves. Much less is known about the sources of carbohydrate in other species such as the various legumes or tree fruits, although with the latter strong correlations between fruit yield and current leaf area have been established.

Table 8.1. Percentage contribution of various parameters to final grain yield. Unpublished data of Dr G. N. Thorne.

	Barley	Awnless wheat
Ear photosynthesis	79	24
Respiration (day)	−24	−28
Respiration (night)	−10	−11
Net ear photosynthesis	79−24−10 = 45	24−28−11 = −15
Transport from flag leaf and sheath	55	115
Total growth	45+55 = 100	115−15 = 100

Supplies of nitrogen, phosphorus and other mineral elements enter the fruit either directly via the roots from the soil or by re-export from older tissues. The amount absorbed from the soil depends on the root growth and the amount of available nutrients remaining in the soil being explored; uptake does occur and responses are obtained to fertilizers applied at the time of ear emergence of cereals, for example. Nevertheless, as mentioned in Chapter 7, there is usually a diminution of root growth with flowering and, under most commercial cropping conditions, mineral nutrients are rarely supplied in excess of the amounts giving economic returns – which must, of course, be physiologically sub-optimal. There is usually appreciable senescence of older leaves so it seems likely that many of the ions entering fruit will be drawn from existing tissues. Moreover, where wheat plants are maintained with a high nutrient concentration surrounding the roots, there is a net loss of nitrogen from the roots, stems and leaves to the developing inflorescence. It would seem that cells, on reaching full size, are unable to retain nitrogen and phosphorus and the supply at sowing of varying amounts of nitrogen to wheat has little effect on the redistribution of nitrogen from the old leaves, about 85 per cent being exported. Removal of the nitrogen supply at anthesis leads to a slight increase in the concentration of nitrogen in the grain because of increased export from the leaves. Whereas the flag leaf is the more important supplier of dry matter to the ear the older leaves are equally or more important suppliers of nitrogen. The higher-order tillers seem to be more of an insurance against the possible effect of damage to the main tillers than of any immediate benefit. They export most of their nitrogen to the flowering regions during senescence, but this amounts to only about 3 per cent of the total amount of nitrogen in the plant. The transport and redistribution of phosphorus probably follows a similar pattern to that described for nitrogen.

A shortage of water during fruit growth leads to reduced size, and hence total yield, probably directly through its effect on the rate of photosynthesis and also by hastening leaf senescence. If the deficit is prolonged, grain weight in cereals is invariably reduced; if it is transient and occurs early in the grain-filling period, the reduction may be compensated by later activity.

There is also a complex interaction between the water relations and mineral nutrition of many fruits and tubers. These organs seem to act as temporary reservoirs of water which are filled at night and during periods of low evaporative demand and are partially depleted when the transpiration rate is high. This leads to the situation where, for large periods of the day, there is either a loss of water from the organ through the xylem or no xylem movement at all. Problems arise, therefore, in the supply of ions such as calcium which, because they cannot move through the phloem, can only enter the organ at night. The result is that there are usually much lower concentrations of calcium in fruit and tubers than in leaves, and this can lead to the onset of physiological disorders associated with calcium deficiency such as blossom-end rot in tomatoes and peppers and bitter pit in apples.

In this chapter, as elsewhere, we have been able to draw attention only to the major features of growth. We have not been able to discuss the very important features of quality, i.e. the type of product required by man. This is a very large and poorly understood topic; the requirements vary enormously between species, involve features related to harvesting and storage such as water concentration in cereal grain, implicate textural characteristics as in potato tubers, and embrace the concentrations and nature of a range of chemical constituents, some of which are simple and others of which have not even been detected.

FURTHER READING

See Chapter 7 (p. 138), plus Evans, L. T. (ed.) (1969). *The induction of flowering*. Macmillan, London.

9

Some Aspects of Overall Growth and its Modification

In the preceding chapters we have followed the growth and development of the component organs, changes in the main physiological processes, and the general relationships of these to the environment in which the plant is grown. As discussed in the concluding chapter, it is unlikely that the growth and resulting yield of any one crop can ever be predicted adequately except by consideration at a level as deep as or deeper than that used in the preceding chapters, and strengthened by a better understanding of the underlying mechanisms. Nevertheless, there are many occasions when information on the growth of the crop as a whole can be most useful. In particular, the identification of times or phases during ontogeny when critical events occur can help in comparing and understanding crop behaviour. The approach used is essentially the reverse of that presented earlier – events are tracked down from the behaviour of the whole, rather than built up from the functioning of the components. Popularly known as 'growth analysis', this procedure has proved most fruitful and has been widely applied since its initiation by F. G. Gregory in 1917.

THE ANALYSIS OF GROWTH CURVES

The dry weight (or energy content) per unit area, Y, or unit plant, W, of any one crop usually follows an approximately S-shaped course, as shown by the points in Fig. 9.1*a*. The difference between two consecutive points of any series gives the average 'crop growth rate' ($\Delta Y/\Delta t$) or the 'absolute growth rate' ($\Delta W/\Delta t$) over that period. During ontogeny, then, there is first a period of accelerating growth rate, followed by a period when it is more or less constant, and then a declining rate – here actually becoming negative in the last weeks because death of leaves and tillers exceeded the new grain growth made. The absolute growth rate may be regarded as a function of two components: the amount of 'growing' material present (number of cells, leaf area, etc.) and the rate of functioning of this as influenced by the environment. If it be assumed that all the growth already made is contributing to new growth, then the logical index to use is the rate of increase of dry matter per unit time per unit of dry matter present. At any instant this, the

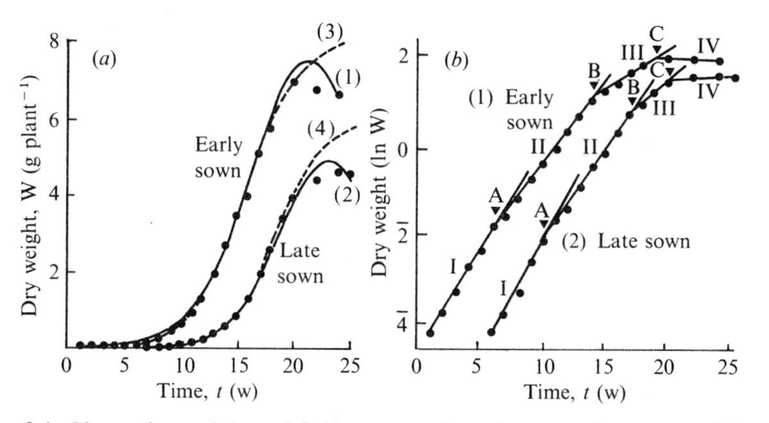

Fig. 9.1. Shoot dry weights of field-grown wheat in an environment of increasing radiation and temperature. Plants described by Equations 2 and 4 were sown 5 weeks later than those described by Equations 1 and 3. The time scale commences 2 weeks after the date of sowing of the first crop and stops at maturity. (*a*) Arithmetic plot, the continuous lines representing the relations

$$(1)\quad W = 0.0154\exp(0.4677t + 0.00048t^2 - 0.000357t^3)$$

$$(2)\quad W = 0.0034\exp\{0.4858(t-5) + 0.00520(t-5)^2 - 0.000695(t-5)^3\}$$

and the broken lines

$$(3)\quad W = 8.10/\{1+\exp(6.636-0.421t)\}$$

$$(4)\quad W = 5.90/\{1+\exp(8.896-0.481t)\}.$$

(*b*) Logarithmic plots which were described by the first two equations above in a continuous form. However, straight lines have been fitted to four growth stages where A, B and C coincide with commencement of secondary root system, cessation of leaf growth and cessation of stem growth, respectively. The relative growth rates, i.e. the slopes of these lines, were

Stage	(1) Early sown	(2) Late sown
I	0.46 w^{-1}	0.52 w^{-1}
II	0.36 w^{-1}	0.40 w^{-1}
III	0.17 w^{-1}	0.27 w^{-1}
IV	−0.01 w^{-1}	0.01 w^{-1}

After Williams, R. F. in Barnard, C. (ed.) (1964), *Grasses and grassland.* Macmillan, London.

'relative growth rate', is given by $(1/W)(dW/dt)$ [$= d\ln W/dt$]. The average relative growth rate over a finite period of time, $t_2 - t_1$, is

$$(\ln W_2 - \ln W_1)/(t_2 - t_1).$$

In a constant environment, the relative growth rate declines throughout ontogeny, mainly because of an increasing proportion of non-dividing to dividing cells. In other words, we can write $dW/dt = W^a$, where a varies from one near emergence to zero as the plant approaches maturity. In

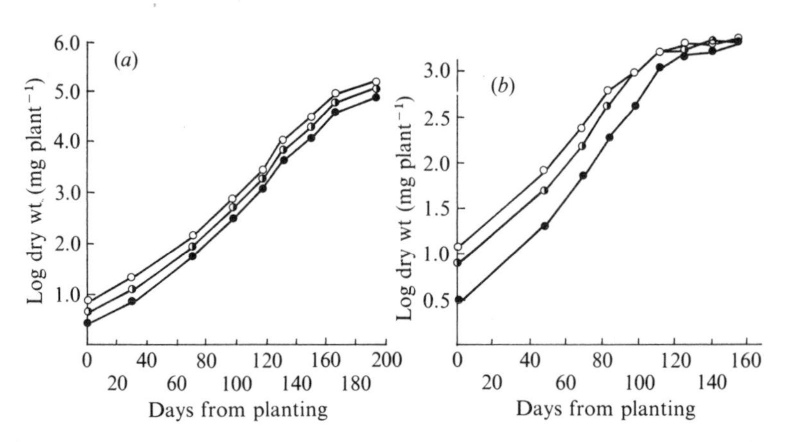

Fig. 9.2. Growth of *Trifolium subterraneum* plants from seeds of 3 (closed circles), 6.4 (pied circles) and 12.5 (open circles) mg mean fresh weight when grown under (*a*) spaced conditions and (*b*) in swards at a density of 618 plants m^{-2}. After Black, J. N. (1957). *Aust. J. agric. Res.* **8**, 335–51.

a natural environment, the same downward drift occurs but this may be compensated by the environment becoming increasingly favourable (cf. Fig. 9.2).

It is obviously convenient when considering growth over a substantial part, or the whole, of the growing season, or when comparing two or more treatments, to express it by a continuous function. This provides a conceptual picture, even if imperfect, of the main features and facilitates comparisons. However, it should be recognized that the very procedure of fitting such a continuous function is a device to extract the general trends and ignore the short-term fluctuations. The latter are based on experimental points and usually represent substantial sampling errors as well as real variations imposed by environmental fluctuations. (The experimenter considers such deviations, of course, and his particular knowledge of the experiment may allow him to identify large divergences resulting from certain conditions.) It is unreal to expect that any such curve will provide a full, clear and biologically satisfying picture of the operation of a complex system over a long length of time and, therefore, it is not necessary to expend a vast effort in seeking a perfect fit. What is much more important is to fit a curve which is adequate in the statistical sense but contains a minimum number of parameters with some biological meaning. Simplicity and interpretability with imperfection are preferable to incomprehensible impeccability.

Two families of curves have proved useful. The first is the polynomial group

$$W = a+bt+ct^2+dt^3+\ldots \qquad 9.1$$

with the minimum number of terms necessary. Frequently, a cubic (as above) is quite adequate and often a quadratic will serve. For statistical and other reasons, it is convenient to fit such expressions in the form

$$\ln W = \ln W_0 + bt + ct^2 + dt^3 + \ldots \qquad 9.2$$

where W_0 is the weight at $t = 0$. The differential of Equation 9.1 describes the change in absolute growth rate with time, and that of Equation 9.2 the relative growth rate. For example, from Equation 9.2

$$(1/W)(dW/dt) = b + 2ct + 3dt^2 \qquad 9.3$$

The other useful family of curves extends from the exponential ($W = be^{kt}$) and time power ($W = bt^k$), where $W \to \infty$ as $t \to \infty$, to a series of asymptotic curves where W approaches a finite value A as $t \to \infty$. Among these is the simple logistic

$$W = A/(1 + be^{-kt}) \qquad 9.4$$

from which the relative growth rate is given by

$$(1/W)(dW/dt) = k(A - W)/A \qquad 9.5$$

This is often useful, especially with isolated organs such as a single leaf (cf. Equation 7.2), but may fail to give a satisfactory fit, as with the data of Fig. 9.1. In the most general form, these curves can be written as

$$W^{1-m} = A^{1-m}(1 \pm be^{-kt}) \qquad 9.6$$

The value of m determines the shape of the curve and in particular the proportion of the final size, A, at which the inflection point occurs. For example, when $m = 0$, there is no inflection point and we have the curve described in Equation 7.2 (p. 128). As m increases, the inflection occurs at increasingly greater proportions of A. Equation 9.4 holds when $m = 2$. Higher values of m should help to accommodate those frequent situations in which growth is near-exponential for an appreciable period and then terminates rather sharply. However, our limited experience suggests that this degree of complexity is rarely warranted, that k often needs to be replaced by a more complex function of time, and even then features such as the decrease in weight brought about by the death of plant parts as in Fig. 9.1 cannot be accommodated. An excellent description of growth curves and their interpretation is given by the late Dr F. J. Richards in Steward (1969).

Rather than force the data to comply with one overall relationship, it may be useful to examine them for a series of discrete functions, as in Fig. 9.1*b*. Although this procedure introduces dangerous elements of subjective judgements and there are unlikely to be abrupt transitions from one phase to another, it has some place when these phases can be identified with definite biological phenomena, as in this example. However, both treatments

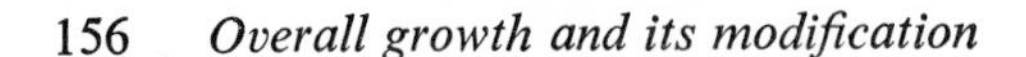

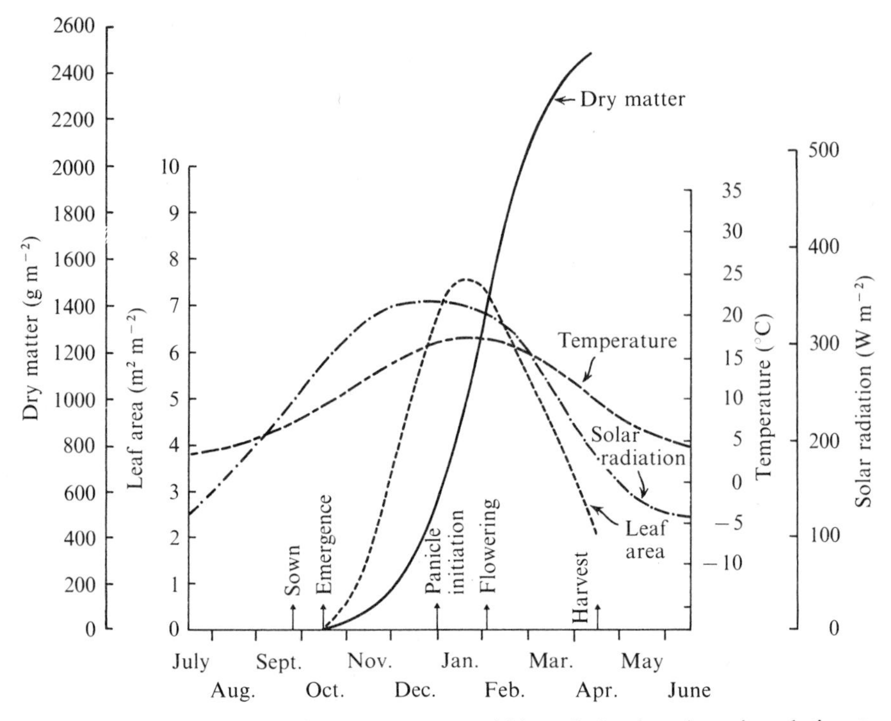

Fig. 9.3. The growth of rice in the Murrumbidgee Irrigation Area in relation to the solar radiation and minimum temperature. After Boerema, E. (1973). *El Riso* **22**, 131–50.

draw attention to the main conclusions of this particular experiment: the falling relative growth rates with time and the higher growth rates of the later-sown crop, which are more than offset by the more rapid development so that it matures at a smaller size. This is particularly so with the phase of inflorescence development (III) where the more rapid growth is associated with a shorter period of differentiation, the production of fewer spikelets, and hence a lower yield.

From Equation 9.2 we see that the weight of a plant (or area of crop) at any time may be expressed as the resultant of three components: (i) its size at the beginning (say, at emergence); (ii) its relative rate of growth; and (iii) the time available for growth. The significance of initial size is often overlooked. When grown as spaced plants an initial difference in embryo size (more exactly, the size at emergence) is often maintained throughout the whole growing season (Fig. 9.2*a*). Here, the increasing relative growth rate with time over the first five months could be ascribed to the increasingly more favourable environment outweighing the downward drift as plant size increases. However, when plants are grown in close proximity to each other, interference between the larger plants results in the growth rate of the more advanced crop being reduced sooner than that of the smaller crop; consequently, all ultimately converge to the same size (Fig. 9.2*b*). This behaviour is discussed more fully in a later section of this chapter.

A growth curve also provides a useful general description of a crop in a particular region; this can be related to general climatic parameters and thereby gives the basic background from which deeper study and analysis can proceed. For example, Fig. 9.3 draws immediate attention to the limitation of the growing season of rice by unfavourable temperatures; the possible growing season is being occupied fully. In descriptions such as these, it is also useful to include leaf area, the commercial harvested component, and the timing of the main developmental phases, such as inflorescence initiation and anthesis.

ONTOGENETIC COURSE OF SIZE AND INTENSITY FACTORS IN PHOTOSYNTHESIS

As some nine-tenths of the dry weight of a plant arises directly from photosynthesis, it is logical to explore effects on growth rates in terms of the size of the photosynthesizing surface (usually taken as the area of green laminae) and the intensity or efficiency at which each unit of leaf functions. We may write

$$\mathrm{dW}/\mathrm{d}t = (1/L)\mathrm{dW}/\mathrm{d}t\,.\,L \qquad 9.7$$

and

$$(1/\mathrm{W})(\mathrm{dW}/\mathrm{d}t) = (1/L)\mathrm{dW}/\mathrm{d}t\,.\,L/\mathrm{W} \qquad 9.8$$

where $(1/L)\mathrm{dW}/\mathrm{d}t$ is known as the 'net assimilation rate', L is the leaf area per plant or per unit soil area (the latter often being called the 'leaf area index') and L/W is called the 'leaf area ratio'. All can be derived from frequent measurements of L and W made throughout the life of the crop. By deriving relationships such as Equations 9.1, 9.4 or 9.6 and differentiating, the absolute growth rate at any point of time is immediately obtained, and this provides directly a good index of the accretion of photosynthate by the plant. Similarly, Equations 9.3, 9·5 or related equations can provide the relative growth rate. The net assimilation rate at each time of harvest is then obtained by dividing the growth rate at that time by the measured leaf area or an estimated leaf area from a curve relating leaf area to time. The average net assimilation rate, E, over the finite time interval, $t_2 - t_1$, is given by

$$\mathrm{E} = (\mathrm{W}_2 - \mathrm{W}_1)/\{\bar{L}(t_2 - t_1)\} \qquad 9.9$$

where the mean leaf area, $\bar{L}$, is obtained by graphical integration, or from fitted curves, or by

$$\mathrm{E} = (\mathrm{W}_2 - \mathrm{W}_1)(\ln L_2 - \ln L_1)/\{(t_2 - t_1)(L_2 - L_1)\} \qquad 9.10$$

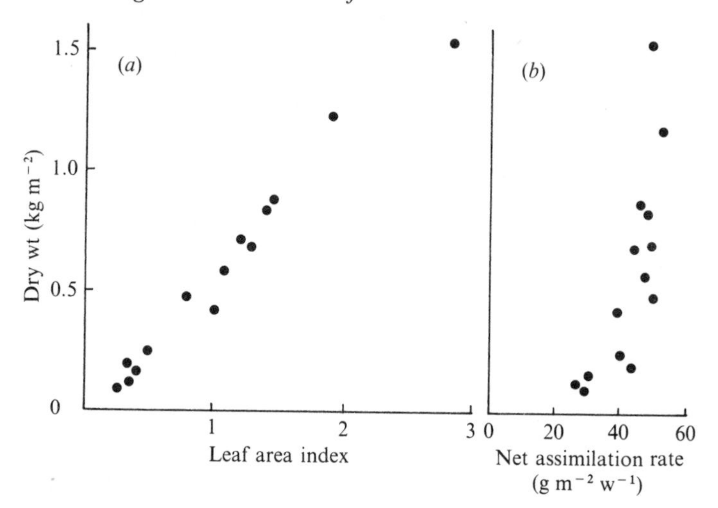

Fig. 9.4. Yield of mangolds as related to leaf area, *L*, and net assimilation rate, E. The plants were grown at Rothamsted with varied fertilizer treatments. Some twenty-fold difference in yield was associated with comparable differences in *L* and a two-fold difference in E. After Watson, D. J. (1947). *Ann. Bot.* **11**, 375–407.

provided that $W = a + bL$ with $b \neq 0$; if these conditions are not satisfied, alternative equations given in the references cited under 'Further Reading' should be used.

A series of investigations along these lines over the last fifty years has contributed greatly to the understanding of crop growth. Perhaps, conceptually, the largest contribution has been to emphasize the efficiency and size components of photosynthesis and to show that under most conditions it is the area of leaf (the proportion of the net photosynthetic gain diverted to leaf) rather than the rate of working of each unit of the photosynthetic machinery which determines the growth rate and the total dry matter yield (Fig. 9.4). This simply illustrates, at the ultimate level of the entire organism, the feature developed in Table 5.2, i.e. in the progression: enzyme content, component reactions, overall process, ... yield, initial differences result in smaller and smaller contributions due to the constraints imposed by other (and often relatively more dominant) effects.

It is often instructive to compare the growth of crops with the potential suitability of the environment for photosynthesis (Fig. 9.5*a*). In this example, it is seen that wheat and barley reach peak leaf area – and hence growth rate – at a time when the net assimilation rate is also maximal. Sugar beet and potato, on the other hand, grow during a period when the environment is becoming increasingly unfavourable for photosynthesis. The diagram immediately suggests that if growth of these two crops commenced earlier in the growing season then higher yields should result; this was indeed

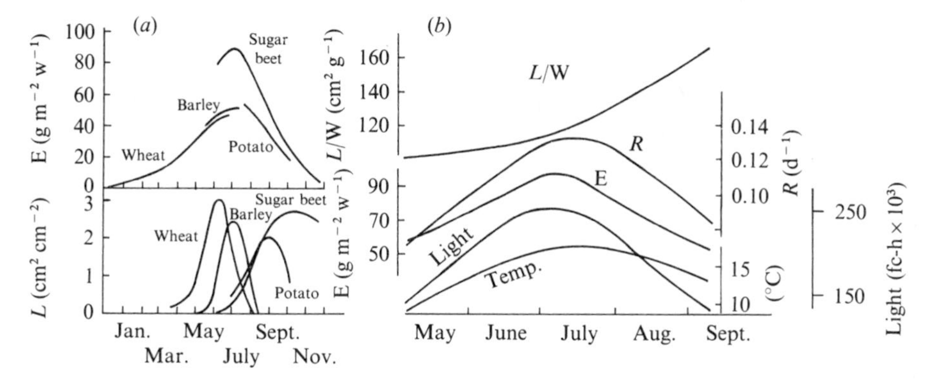

Fig. 9.5. (*a*) Smoothed curves showing the average changes with time in net assimilation rate, E, and leaf area, *L*, of four crops grown at Rothamsted, 1934–43. After Watson, D. J. (1947). *Ann. Bot.* **11**, 41–76. (*b*) The relative growth rate, *R*, net assimilation rate, E, and leaf area ratio, *L*/W, of young sunflower plants of comparable size grown at Oxford (similar climate to Rothamsted) at different times of the year, together with the average temperature and total luminous flux density (in photometric units) during each experimental period. After Blackman, G. E. in Zarrow, M. X. (ed.) (1961). *Growth in living systems.* Basic Books, New York; cf. also *Ann. Bot.* (1955) **19**, 527–49.

found to be so, provided the constraints which determined the traditional growing season were overcome. (In sugar beet this was premature flowering following vernalization and was met by selection; in potato, it was susceptibility to late frosts and could be partially met by earlier planting and also by increasing the amount of growth made before planting.) The net assimilation rates shown in Fig. 9.5*a* include a component ascribable to the environment and another representing an inherent downwards drift with time. The effect of the time drift can be eliminated from experiments by planting successive crops at short intervals throughout the growing period and measuring the growth parameters of plants initially of the same size over a period of, say, one week (Fig. 9.5*b*). Possibly the most useful index of growth is the relative growth rate, although the net assimilation rates given in Fig. 9.5*a* may be compared directly with those in Fig. 9.5*b* despite the difference in species. It will be noted that the variations in relative growth rate involved changes in both its components, E and *L*/W, and that neither of these was determined solely by light and/or temperature. Actually only about 51, 67 and 39 per cent of the variation in relative growth rate, net assimilation rate and leaf area ratio, respectively, could be accounted for by multiple regression on the total visible radiation received and the mean temperature.

Much of the early work suggested that the net assimilation rate remained fairly independent of internal influences, variations being ascribable entirely to environmental factors. (Of these, the aerial environment also seemed to be predominant as large differences in mineral nutrition had little effect.) If these initial results had been confirmed by further work, the interpretation of all the complicated interrelationships would have been relatively simple since they would have been dependent only on the leaf area ratio. The net assimilation rate has, however, been shown to decrease with age. The decrease is caused by the shading of lower by upper leaves, the decreasing photosynthetic capacity of later formed leaves and effects of 'sinks' elsewhere in the plant on photosynthesis. In addition, there is an increasing proportion of respiring to photosynthetic tissue and the respiration rate is related to the rate of photosynthesis. The net assimilation rate and leaf area ratio have not proved, therefore, to be particularly valuable, even as intermediate parameters, in understanding the responses of crops; it is usually better to proceed directly from the descriptive stage reflected by absolute and relative growth rates to a more detailed analysis.

ONTOGENETIC COURSE OF MINERAL NUTRITION

Although most of the substance made by a crop during its growth comes from photosynthesis, the total production achieved is frequently a function of the supply of mineral nutrients. These constitute only a very small proportion of the dry weight; nitrogen, phosphorus and potassium being usually about 1.5, 0.2 and 1 per cent respectively, and micronutrients even less. For example, molybdenum can be as low as 10^{-7} per cent. With the possible exception of water, supplies of the mineral elements are probably the most readily manipulated of all environmental factors. In agricultural situations, however, they are rarely provided at concentrations greater than those which give the maximum economic return. These concentrations are usually less than those which would allow the plants to exploit fully all the other environmental variables, and hence less than the needs for maximum productivity. Economic and management constraints also usually dictate that the nutrients are supplied at sowing, although sometimes additional amounts may be supplied later.

The roots of the young seedling penetrate soil with a relatively high nutrient concentration. As discussed in Chapter 4, the rate of nutrient uptake is proportional to the concentration in the soil solution and to the extent of the root system, with a proportionality constant dependent on the species of plant and ion in question. If the external concentration is sufficiently high, the rate of uptake more than compensates for the rate of use in the growing centres and nutrients will accumulate within the plant. A high internal concentration, especially of nitrogen, in turn stimulates the development of new meristems (e.g. Fig. 7.6) and so increases the demand for more mineral ions. This is partially met by the expanding root system exploring

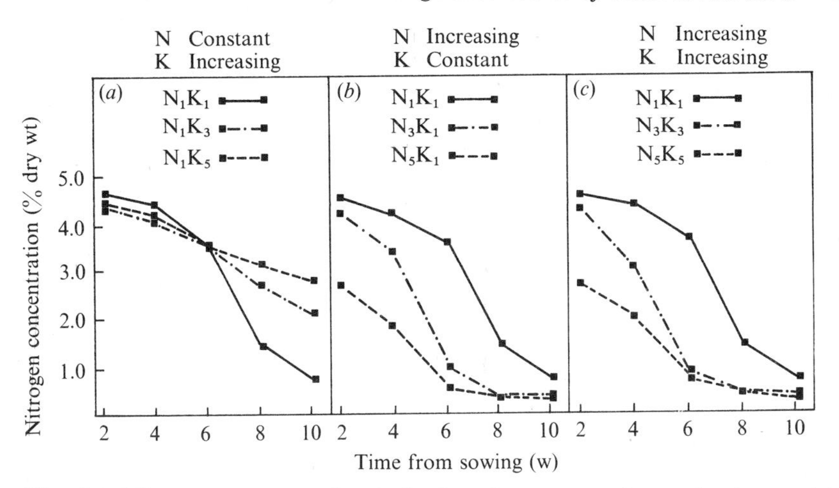

Fig. 9.6. Nitrogen concentration in barley plants grown in sand culture with all nutrients added at sowing. $N_1:K_1$ was 3.6:1, subscripts 1, 3 and 5 denote highest concentration and 1/9 and 1/81 of that concentration, respectively. Therefore, (*a*) indicates a series with K minimum and N in excess, (*b*) N minimum and K in excess, and (*c*) a balanced series. After Gregory, F. G. (1937). *A. Rev. Biochem.* **6**, 558–78.

previously untapped volumes of soil, but the concentration around the existing roots falls. The replenishment from organic matter and soil colloids never keeps pace with root uptake and immobilization reactions in the soil, and root expansion gradually falls behind shoot growth; consequently, the products of photosynthesis accumulate faster than mineral ions are absorbed. The concentration of a mineral nutrient (amount per unit dry weight) is therefore high in a seedling soon after emergence, may change little for some time, but eventually decreases (Fig. 9.6).

It seems likely that with most crops a stage is reached where the demand from new growing centres overtakes the supply to the plant as a whole; this is known as the stage of 'internal starvation'. It is a useful general concept indicating the balance between demand and supply of nutrients; indeed, in one sense, the most efficient management would be one in which the plant was always kept at the point of onset of internal starvation. However, when based on the whole plant, it loses some of its value because of the recycling of nutrients from the older leaves. The plant meets current growth requirements by a combination of uptake and transfer from older tissue, the relative importance of the two processes being dependent on the rate of growth and the availability of nutrients in the soil. The penalty paid for this transfer is the more rapid senescence of older leaves; if these are heavily shaded by younger leaves their photosynthetic contribution is likely to be small and their main contribution to the growth of the plant may be through the ions they supply. It seems likely that even an appreciable degree of internal

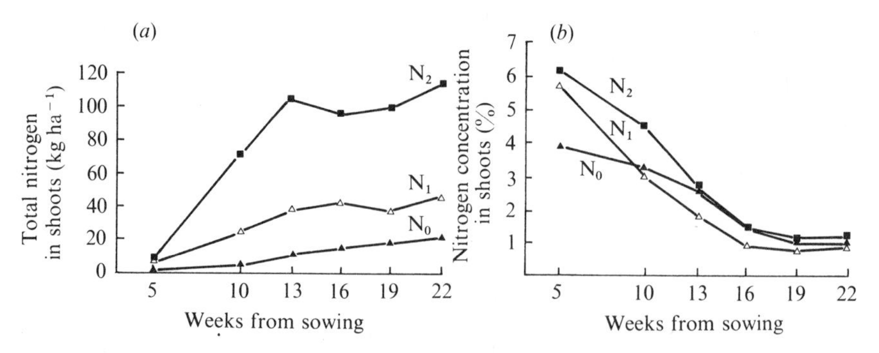

Fig. 9.7. Uptake of nitrogen by and concentration in shoots of a wheat crop grown in Western Australia. Treatments were N_0, no applied nitrogen; N_1, 56 kg N ha^{-1} applied at sowing; and N_2, 112 kg N ha^{-1} applied at sowing, 5 weeks after sowing and 10 weeks after sowing (i.e. a total of 336 kg). The dry weights of shoots at the final harvest were 2400, 5800 and 9400 kg ha^{-1} and the concentrations of N in the grain 1.82, 1.72 and 2.60 %. After Halse, N. J., Greenwood, E. A. N., Lapins, P. & Boundy, C. A. P. (1969). *Aust. J. agric. Res.* **20**, 987–98.

starvation may not reduce yield or retard growth greatly, but to explore this quantitatively would need a detailed analysis beyond our present terms of reference.

It follows from the above that the concentration of a mineral in a seedling reflects fairly exactly the concentration in the soil solution (Fig. 9.6*b*, *c*). Provided all other nutrients are adequate, this initial wide range of possible concentrations converges to very much the same low level as the plant approaches maturity. However, if the supply of nutrients other than the one being considered is sufficiently low to restrict plant growth to a greater extent than the uptake of the nutrient in question, the concentration of the latter may remain high (cf. Fig. 9.6*a*). In this situation there is what is termed 'luxury consumption' of an ion, i.e. greater concentrations are absorbed than can be used. One should always aim to grow plants with a 'balanced' soil solution. The best index of this is given by the ratios of the elements to each other in the fruit; this ratio (but not of course the concentration) appears to be independent of the degree of unbalance in the soil solution in which the plant was grown. Nutrients in excess remain in the leaves, not being translocated to the fruit. The balanced ratio probably varies with species and possibly with cultivars; in one variety of barley the ratio of N:K:P was found to be approximately 13:4:1 and in the banana was 8:16:1.

Possibly, most empirical fertilizer trials tend to integrate the aspects outlined above, indicating the optimum rates of application to give the best economic return (or even highest productivity) for the variety used at a particular site in a particular season. They reflect (but rarely elucidate) the operation of features such as initial concentrations and rates of release and immobilization in the soil, rates of root expansion and of photosynthesis in the various leaf layers, the amount of transport from older leaves to the current growing organs, and so on. The fact that the degree of the overall response varies quite markedly from season to season with the one crop at a particular site suggests that a number of the components of this intricate interrelationship influence the end result, but it has not been possible yet to describe these in any formal rigorous fashion. Most field crops do continue to import some mineral ions from the soil until maturity (Fig. 9.7) and the ratios of these to dry matter always decrease during ontogeny; nevertheless, it is quite uncertain whether it is the amounts of nutrients that can be absorbed or the net amounts of photosynthate that can be made which impose the ultimate limitation on production. Indeed, it is equally uncertain which of these basic supply factors is primarily involved in the intermediate yield-determining responses, such as cessation of branching or abscission of flowers and fruits; it is also possible, though unlikely, that responses in which hormonal mechanisms feature prominently may be completely independent of the rates of operation and balance between the two basic supply processes.

The supply of mineral nutrients also needs to be adjusted to certain intrinsic genetic and seasonal features. For example, the high-yielding cereals maize and rice maintain some leaf area beyond the time when the grain is fully grown, whereas the lower-yielding wheat and barley usually experience complete senescence of leaves before the end of grain growth, thereby preventing some of the yield potential from being realized. With the latter, one aim of fertilizer use is to maintain the leaf surface for as long as possible. In the potato, tuber initiation is delayed by extensive shoot growth induced by the liberal use of fertilizers, but very high yields would be achieved ultimately if the growing season could be prolonged sufficiently. In regions such as Europe, where the season is restricted by low temperature and frosts, this potential cannot be reached and a compromise must be effected between earliness of initiation, encouraged by low nutrient supply, and the rate and duration of tuber growth and persistence of leaves, encouraged by a high supply. Similar considerations apply to the growth in south-eastern Australia of cotton – a crop which flowers and fruits over a prolonged period.

Many attempts have been made to derive relationships between the final harvested yield (of total dry matter, all-above-ground portions, or of the commercial product) and the amount of fertilizer applied. Possibly, the simplest and most widely used is that first proposed by Mitscherlich, in which it is assumed that when the yield, Y, is plotted against the amount of a nutrient (say, nitrogen) applied, N, the degree of increase of Y with N at any point is proportional to the amount by which the yield falls below an upper limit, A, this being determined by all the other growth factors. That is, $dY/dX = a(A-N)$, which on integration gives

$$Y = A\{1-\exp(-aN)\} \qquad 9.11$$

This is, of course, the curve describing the course of a monomolecular reaction, and expresses the 'law of diminishing returns'; it has already been used in Equation 6.3. This relationship holds reasonably well when all factors except the one being studied are maintained at their optimal levels; however, it cannot be conveniently extended to cover the variation of two or more nutrients or the decrease which often occurs at very high concentrations. If two or more nutrients are increased together, a linear or a sigmoid curve is often found. In other words, the extension in the form

$$Y = A\{1-\exp(-a_N N)\}\{1-\exp(-a_P P)\}\{1-\exp(-a_K K)\} \qquad 9.12$$

is not supported by experimental observations.

Another approach is that of the 'resistance' formula in which the reciprocal of yield is expressed as the sum of independent functions for the different growth factors; i.e.

$$1/Y = f'(N)+f''(P)+f'''(K)+\ldots+b \qquad 9.13$$

One variant used with some success is the form

$$1/Y = a_N/(N_0+N_f)+a_P/(P_0+P_f)+a_K/(K_0+K_f)+b \qquad 9.14$$

where the constants a_N, a_P, a_K are specific for the nutrient and crop and the subscripts 0 and f indicate the amounts initially present in soil and seed and that added in the fertilizer, respectively. A further variant is that fitted to the data of Fig. 9.8; this was used in the form

$$1/Y = [1/\{1-(N_0+N_f)/b_N\}][1/A+1/\{a_N(N_0+N_f)\} + 1/\{a_P(P_0+P_f)\}+1/\{a_K(K_0+K_f)\}] \qquad 9.15$$

where all symbols have the same meaning as before and b_N is the amount of nitrogen in the soil which stops growth through osmotic effects. (The amounts and concentrations are such that a given weight of nitrogen is likely to have much larger osmotic effects than similar weights of potassium and phosphorus.) This gave an acceptable fit to observations from a particular crop in a given environment and allowed a satisfactory estimation of the amounts initially present. However, the values of the various a constants varied appreciably between species, which is acceptable, and between sites and seasons, which is not. Although worth much further exploration, simple functions of this nature must necessarily be limited in application, and can hardly be expected to relate widely separated components of a complex system with a high degree of precision.

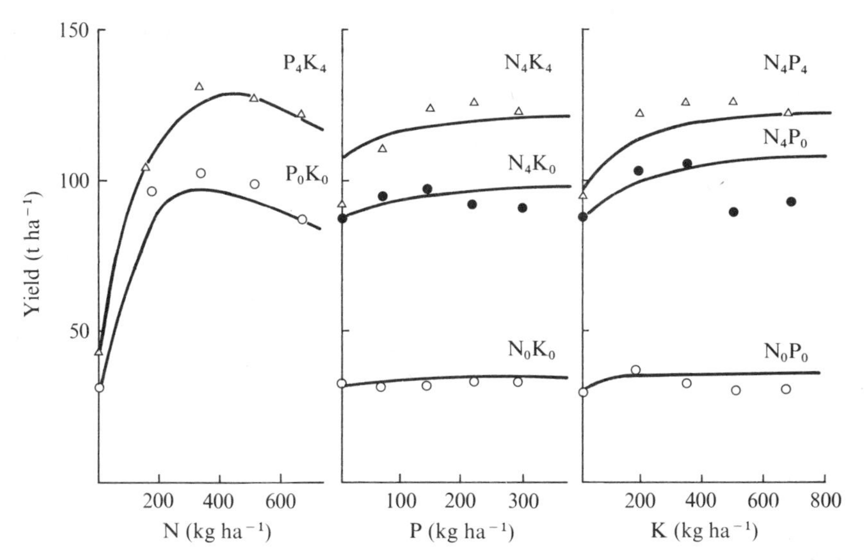

Fig. 9.8. Relationships between yield of cabbage and amounts of applied nitrogen, phosphorus and potassium in an experiment carried out in the Midlands region of England. The subscript 0 indicates no application and 4 the highest rate of application. Results for the extreme treatments only are shown. The curves were calculated from Equation 9.15. After Greenwood, D. J., Wood, J. T., Cleaver, T. J. & Hunt, J. (1971). *J. agric. Sci., Camb.* **77**, 511–23.

ALLOMETRIC AND OTHER INTERRELATIONSHIPS DURING GROWTH

A weight of evidence shows that a plant is a very closely integrated organism; even the fact that members of a given cultivar are always recognizable as such shows a large degree of stability. (Indeed, if this were not so, classification on morphological characters would be impossible.) Yet one individual varies in size over several orders of magnitude during ontogeny and there are often large differences in size between members of like genetic constitution grown in different environments. The question then arises as to the degree of proportionality maintained between the various organs under these conditions. If this stays constant or is described by tractable relationships, then the analysis and prediction of even complex events are greatly facilitated.

One relationship which has been found to have reasonably wide application is the simple allometric equation

$$Z = bX^a \quad \text{or} \quad \log Z = \log b + a \log X \qquad 9.16$$

where Z and X represent the sizes of two variables. This may often hold at organ level, describing the relationship between two of its dimensions during development, e.g. length to breadth or area of a leaf, length to

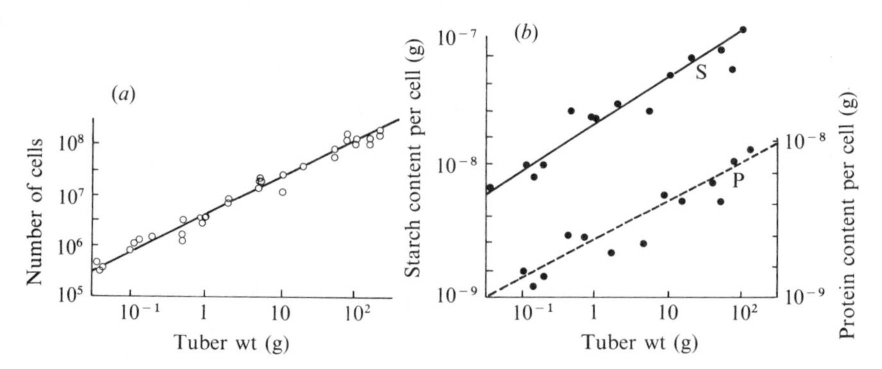

Fig. 9.9. Relationships of (*a*) number of cells per tuber, N, and (*b*) starch, S, and protein, P, contents per cell with potato tuber fresh weight, W_F. The relationships are described by:

$$N = 3.71 \times 10^6\, W_F^{0.73}; \quad S = 2.0 \times 10^{-8}\, W_F^{0.35}; \quad P = 2.1 \times 10^{-9}\, W_F^{0.24}.$$

After Plaisted, P. H. (1957). *Pl. Physiol.* **32**, 445–53.

diameter of a fruit, number of xylem cells to leaf length, contents of various substances in developing organs (Fig. 9.9), etc. The constant b is simply the value of Z when X is unity; here, for example, it shows that a tuber of 1 g contains 3.7×10^6 cells. The constant a indicates the degree of proportionality between the two variables. Differentiation of Equation 9.16 with respect to time gives

$$(1/Z)dZ/dt = a\{(1/X)dX/dt\} \qquad 9.17$$

If $a = 1$, then the relative growth rates of the two variables are equal (i.e. the absolute growth rates are proportional to each other). It also follows that where X and Z represent two dimensions of an organ and $a = 1$ the organ does not change its shape as it grows; if $a \neq 1$ the shape does change during growth.

In those situations where the ranges of X and Z are only about one order of magnitude, it may not be possible statistically to distinguish between Equation 9.16 and the linear relationship $Z = b + aX$; that is, it will be difficult to decide whether the relative or absolute rates maintain proportionality and equally good predictions could be given by either relationship.

When we proceed from an individual organ to a group of organs, then it is rarely found that the relationship holds in this simple form. Nevertheless, in some species, a good description is given by the more complex variant

$$Z = bX^{(a + c\log X)} \quad \text{or} \quad \log Z = \log b + a\log X + c(\log X)^2 \qquad 9.18$$

For example, the relationship between root and shoot weight in sugar beet is very constant, the root comprising an increasing proportion with ontogeny (Fig. 9.10). This pattern is not changed by large variations in irradiance,

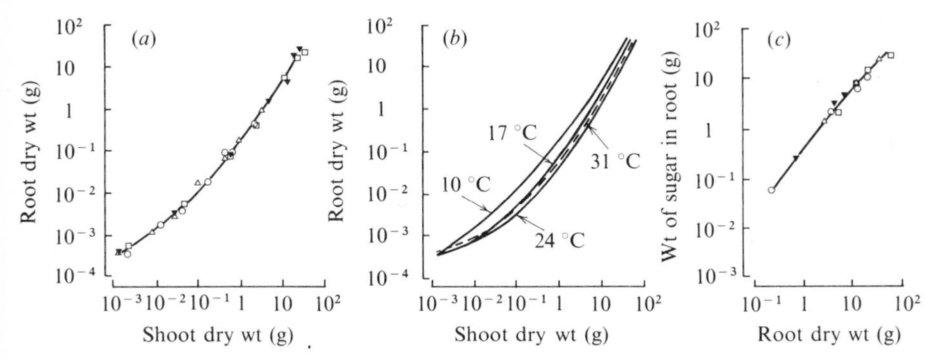

Fig. 9.10. Allometric relationships between dry weights of storage root and shoot of sugar beet when grown (*a*) with different flux densities of visible radiation, and (*b*) at different temperatures; (*c*) shows the relationship between weight of sugar and weight of storage root. After Terry, N. (1968). *J. exp. Bot.* **19**, 795–811.

although the rate of growth is greatly influenced. However, the pattern is changed by differences in temperature (and nitrogen supply), a higher proportion of material being diverted to the root at low than high temperatures (and at low than high concentrations of nitrogen in the soil solution). It was also found in this study that the amount of sugar in the storage root was closely related to the dry weight of the root irrespective of irradiance or temperature.

Relationships with this degree of constancy are invaluable in predicting the behaviour of the crop. Given that the growth of the whole in a particular environment can be estimated – an issue discussed in Chapter 10 – then most of the other necessary entities, such as yield of sugar, can be obtained easily. When considering a crop they should always be explored and at least the degree of variation of the parameters induced by the different environmental factors ascertained. The exercise itself gives a much better picture of the responses of the crop and leads to clearer understanding.

However, it must be recognized that with many crop species the degree of constancy in the relationships is too small for them to be of more than descriptive value. Some general issues may be mentioned in respect of three so-called 'root' crops. Following the procedures discussed in an earlier section of this chapter, we may describe the growth of the crops by the ontogenetic courses of leaf area and the harvested product in 'average' situations (Fig. 9.11). In the potato, there is first a development of the leaf surface. Following the initiation and growth of tubers, new leaves cease to expand and the existing leaves senesce rapidly; tuber growth decreases when the leaf area falls below about 1. With different cultural conditions, the slope and position of the curve describing tuber growth relative to leaf area can be displaced to an appreciable degree. If displaced to the left, i.e.

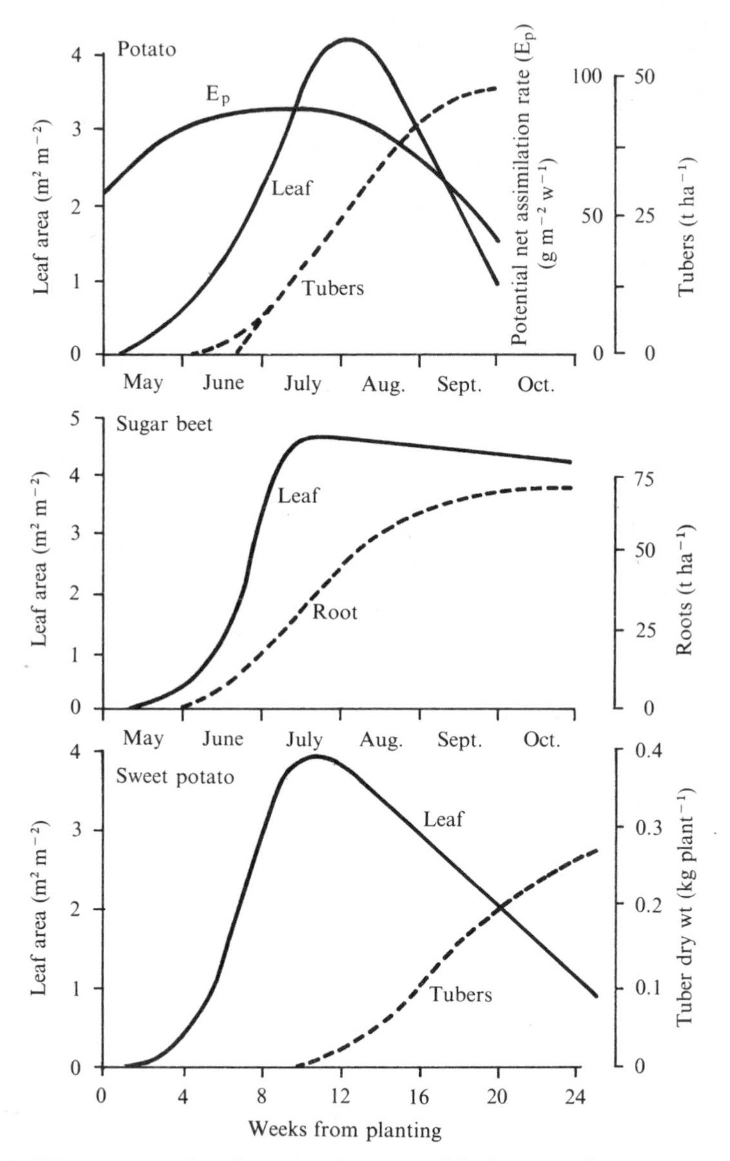

Fig. 9.11. Diagram showing the general course of leaf area and storage organs of (*a*) potato and (*b*) sugar beet in the Midlands region of England, and (*c*) of sweet potato in the West Indies. E_p indicates the potential net assimilation rate as defined in Fig. 9.5*b*. Data provided by Drs R. K. Scott and C. J. Walter feature prominently in (*b*) and (*c*), respectively. After Milthorpe, F. L. (1969) in *Proc. Int. Symp. Tropical Root Crops, Trinidad*, vol. 1.

tubers formed on plants with a small leaf area, the rate of growth of tubers is low and the final yield achieved small. If displaced to the right, the rate of growth of tubers is high and continues for a longer time, provided the leaf area is not destroyed by frost or blight. Growth of sugar beet is much more rigid, and although the slopes of the leaf area and root growth curves can be altered in both position and slope relative to time, they do not change much relative to each other. Sweet potato seems to be somewhat intermediate between the two, but not much information is available for behaviour in a wide range of environments. In this example, tuber growth would seem to be delayed relative to leaf growth much more than is desirable for high yields. A number of questions immediately arise, such as the apparent late initiation of tubers, slow growth rates and rapid senescence of leaves. These need exploring more deeply.

SPATIAL RELATIONSHIPS OF CROPS

Most crops consist of populations of individuals with like genetic constitution, commencing growth at much the same time from propagules of much the same size, and more or less uniformly distributed over an area of soil. Usually, the plants are at first sufficiently widely separated so that they do not interfere with each other, but eventually each plant modifies the environment of adjacent plants making it less favourable for growth, and the growth rate falls below that of isolated plants. The departure of the growth rate from that of completely isolated plants is a function of the proximity of roots and shoots of adjacent plants, and therefore depends primarily on the initial density and the amount of growth made. As seen earlier (Fig. 9.2), the initial embryo size has an appreciable effect on the time at which mutual interference commences; equally, differences in relative growth rate have a large influence. Crop plants are almost invariably grown at densities which lead, early in ontogeny, to substantial interference – and, hence, a reduction in the rates of growth below those of isolated plants in the same environment. The resulting responses give interrelationships between yield, Y (weight per unit area), and density, d (number of plants per unit area), varying with time in the manner illustrated in Fig. 9.12. These relationships can be usefully described by

$$\mathrm{Y} \doteqdot bd^{a}, \quad \mathrm{W} \doteqdot bd^{a-1} \qquad 9.19$$

where W is the weight per plant (note $\mathrm{Y} = \mathrm{W}d$). Here a and b are parameters which are constant at any one time but vary with time. At planting, $a = 1$; i.e. the weight per plant is independent of density, and yield is proportional to density. As the crop grows, a departs from unity at progressively lower densities; this point is a measure of the density at which interference becomes measurable. Over the range in which there is interference, a is usually constant

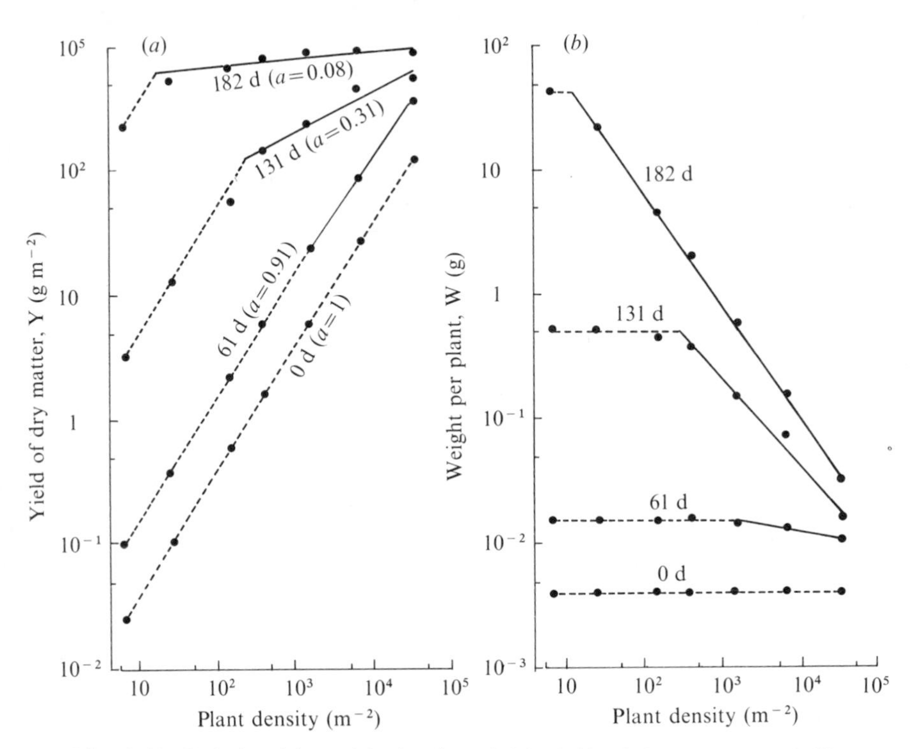

Fig. 9.12. Relationships with density of (*a*) yield of dry matter and (*b*) average weight of plants of subterranean clover at 0, 61, 131 and 182 days from sowing. The densities ranged from 6 to 34000 plants m^{-2}. Interference between plants (indicated by continuous lines) was detected at densities of 1500, 280 and 13 plants m^{-2} on days 61, 131 and 182, respectively. Broken lines indicate ranges over which there was no interference (a = 1). Redrawn from data of Donald, C. M. (1951). *Aust. J. agr. Res.* **2**, 355–76; cf. also Kira, T., Ogawa, H. & Sakazaki, N. (1953). *J. Inst. Polytech. Osaka City Univ.* **4**, 1–16.

at any time but it falls with time. Consequently, late in development the situation is often reached where a falls to zero over the entire range of d; yield is then independent of density and the weight of each plant is inversely proportional to density. Growth, however, rarely ceases at this stage, both dY/dt and dW/dt remaining positive. The parameter b has no useful biological meaning; mathematically it indicates the yield at unit density provided the relationship holds over this range, but this is rarely so. The relative growth rates of the plants not showing interference provide a useful measure of the effect of the unmodified environment.

The 'plateau' (a = 0) of the relationship between total dry matter and density is usually maintained over a range of densities of several orders of

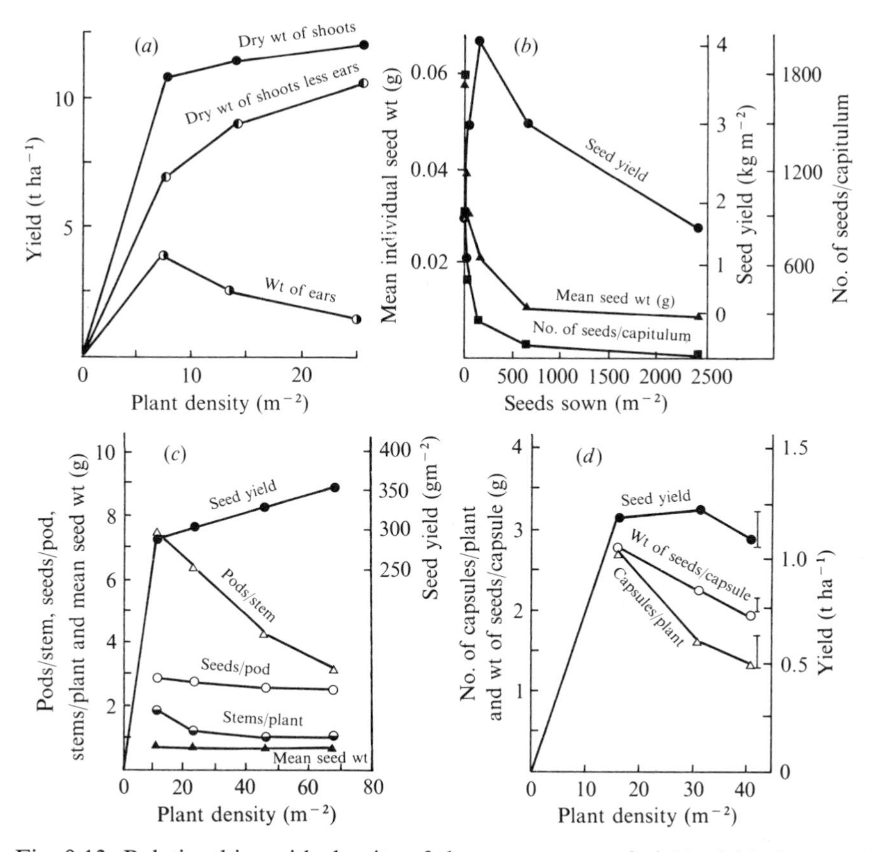

Fig. 9.13. Relationships with density of the components of yield of (*a*) shoot and ear dry weight of maize, (*b*) sunflower (there may have been some plant death, hence the plant population may have been less than the number of seeds sown), (*c*) field bean, and (*d*) oil-seed poppy. After Harper, J. L. (1961). *Symp. Soc. exp. Biol.* **15**, 1–39, where sources of the data are acknowledged.

magnitude. This is usually not so with yield of seeds, fruits or specific organs such as storage roots and tubers; here, there is usually an intermediate density at which the yield is maximal (Fig. 9.13). An appreciable number of experiments indicate that, at least with cereals, the minimum density giving the plateau yield of dry matter (i.e. the density at which $a = 0$) is also the density at which grain yield is maximum. This is approximately so with the data of Fig. 9.13*a*. These responses arise because the effects of increasing numbers of plants per unit area are more than offset by the reduced yield of seed (or fruit) per plant. The latter is always ascribable to a reduced number of seeds per plant, N; i.e.

$$\mathrm{N} \doteq c\mathrm{W}^{k} \qquad 9.20$$

where c and k are constants. (This relationship holds more exactly if W represents the weight of a fruit-bearing axis, such as a tiller of a cereal, rather than the weight of the whole plant.) Here, N reflects differences in both the number of inflorescences produced and the number of seeds formed per inflorescence. Where only one inflorescence is formed per plant (as in the data of Fig. 9.13*b*), there is also an appreciable reduction in the average weight of a seed; where many inflorescences are formed, particularly over a long time span, the average weight per seed is remarkably constant for any one cultivar (Fig. 9.13*c*, *d*). (See also Fig. 8.5.)

The above idealized relationships are usually achieved only if seeds of uniform size are sown with uniform spacing and all plants emerge at the same time. This is rarely observed. Usually, the larger plants at emergence, arising from larger seeds or better germination conditions, dominate smaller adjacent plants, this effect becoming more and more accentuated with time. The relative degree of variation between all plants in a stand therefore increases with time and a number of the smallest plants may die without setting seed. For example, extending the data of Fig. 9.2 to the situation where seeds of the different sizes were sown together, it was found that eventually the same total yields were obtained but that this was due almost entirely to plants arising from the larger seed. The same considerations apply to mixed communities, success usually going to that species which attains the largest size earliest in the season.

These effects arise from interference with the light, nutrient and water supplies to the individual plants. There is never any shortage of space as such; even with a very dense crop (say, $L \approx 10$), only about 1 per cent of the total canopy volume is occupied by plant parts and in a soil layer with $R_V = 20\ \text{cm}^{-2}$ about 2 per cent of the pore space by roots. The rates of movement of carbon dioxide through the crop are such that the concentration is never seriously depleted in any layer. Temperature gradients within the canopy are not greatly influenced by differences in density. As with changes in time at any density, increase in density results in a larger leaf area and a decrease in net assimilation rate at any one time; the subsequent effect on the crop growth rate depends on the relative changes in and contributions by each of these two components. Fig. 9.14*a* gives a fairly reasonable overall picture of these, although the positions of the curves vary with time and species. For instance, the increase in crop growth rate with density is likely to be found only at low densities with young stands; moreover, it frequently declines more over the higher range, especially in older stands. Although the leaf area curve frequently decreases, especially in older stands, it is rare for the net assimilation rate, and hence the crop growth rate, ever to fall to zero. The lower assimilation rates at high densities lead to root systems proportionately smaller than the above-ground parts of the plant and thereby to a smaller amount of nutrients being absorbed (Fig. 9.14*b*). This no doubt

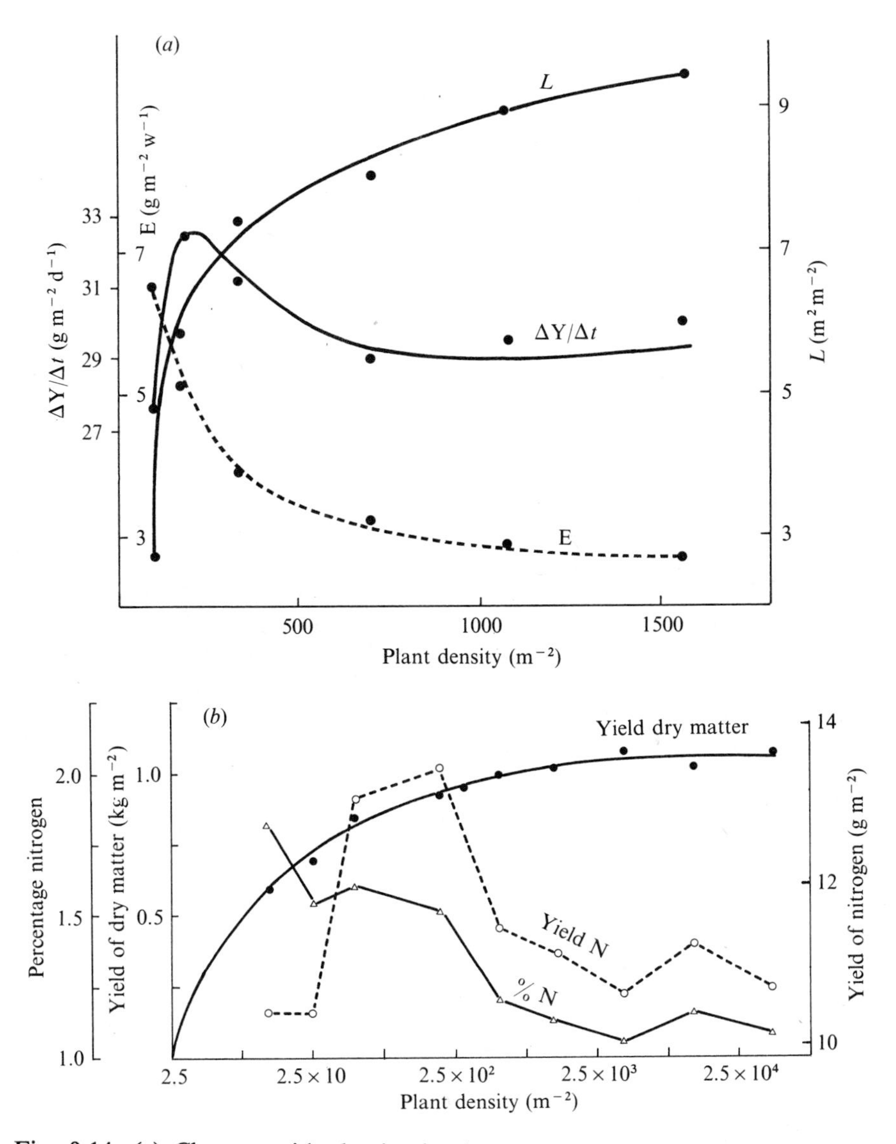

Fig. 9.14. (*a*) Changes with density in the crop growth rate, $\Delta Y/\Delta t$, the net assimilation rate, E, and the leaf area of *Brassica oleracea* at the same time from sowing. After Blackman, G. E. in Eckardt (1968).

(*b*) The total yield of above-ground shoots and content and concentration of nitrogen in swards of Wimmera rye grass (*Lolium rigidum*) grown at the densities shown. After Donald, C. M. (1951). *Aust. J. agric. Res.* **2**, 355–76. N.B. In both figures not all arithmetic scales start from zero.

contributes to a decreased rate of leaf expansion and the decrease in crop growth rate, although there is some compensation from recycling within the plant. The main effect of water is through the smaller root systems of the individual plants at high densities and thereby the smaller reservoir of water available in each cycle of rewetting. Therefore, at high densities, the crop needs to be watered more frequently than at low densities if it is not to experience a water deficit. It will be appreciated that the above features are closely interwoven and that it is not possible to describe them quantitatively without formulating a detailed interdependent system.

CHEMICAL MODIFICATION OF GROWTH

As stated in the beginning, we have deliberately ignored growth substances and other mechanisms of stabilizing and integrating the growth and development of a plant. They are undoubtedly concerned in this integration, but the ways in which they are involved still remain largely speculative and fuller treatment would not have helped us greatly in resolving many of the empirical relationships we have had to use. Nevertheless, this chapter should not be closed without some mention, albeit cursory, of the modification of growth by chemicals, particularly growth substances, currently used in agriculture.

Possibly the most successful application has been their use in a destructive sense – in facilitating the removal of unwanted species in a mixed community. A large number of selective herbicides have been developed with a wide range of degrees of selectivity and methods of action. The most widely used are the auxin-type herbicides such as the several formulations of 2,4-dichlorophenoxyacetic acid; its mode of action is possibly by activating indiscriminately a large number of otherwise repressed genes and its selectivity for different species resides in differences in absorption, transport and breakdown in the plant. However, our concern is more with the application of substances to regulate the growth of crops. Again, synthetic auxins have been widely used in a number of transient control processes, such as controlling the number of fruit set, preventing abscission, regulating the onset of flowering, stimulating root formation on cuttings, and so on. In this application it seems that the most effective compounds are synthetic analogues, with sufficient similarity to the native substances to participate in the required reaction but sufficiently dissimilar to be only slowly inactivated by the control systems regulating the endogenous auxin. Gibberellins are used to stimulate germination during malting and to increase fruit set and bunch quality of seedless grapes; various growth retardants are used to produce a desired type of growth of chrysanthemums and other ornamental species and to regulate fruit retention on apple trees, and ethylene is used to regulate the ripening of banana fruits. Costs of application and the transient nature

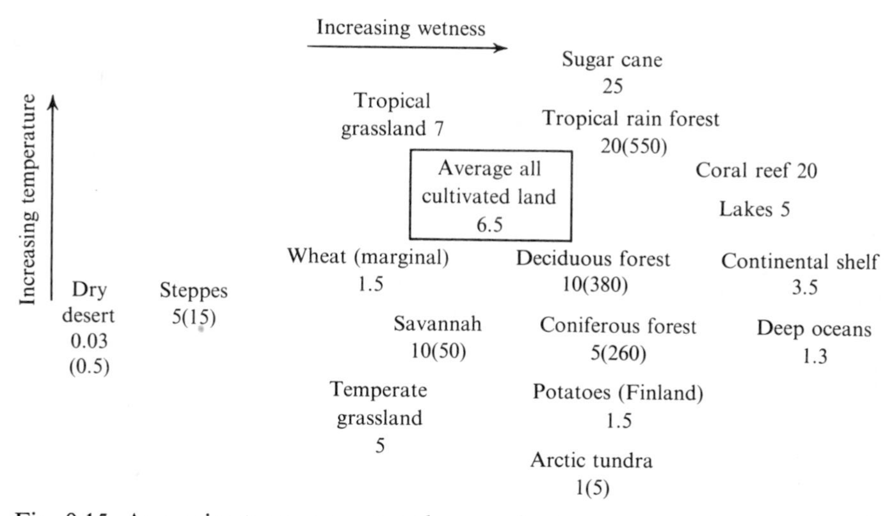

Fig. 9.15. Approximate average annual rates of net primary production of total dry matter (t ha^{-1} y^{-1}) and estimates of the mass of standing material (t ha^{-1}) of a number of natural ecosystems and of agricultural crops in comparable geographic zones. Data after Rodin, L. E. & Basilevič, N. I. in Eckardt (1968). Loomis *et al.* (1971), Lieth, H. (1972). *Nature and Resources* **8**, 5–10, and other sources.

of the effects limit approaches to high-return crops and to such operations where rigorous control of a developmental phenomenon such as flower initiation or fruit abscission is required. Quality can be improved and management procedures facilitated, but it will be a long time before it is economically feasible – even if it is physiologically feasible – to use growth regulators in increasing the productivity and efficiency of large-scale agricultural crops.

PRODUCTIVITY AND THE INCREASE OF YIELD

In concluding this chapter, it may be of interest to examine the productivity actually achieved by agriculture. In both natural and agricultural ecosystems (Fig. 9.15) this is determined largely by the length and constancy of the growing season, which in turn is determined either by temperature or the supply of water. For example, H. Lieth finds that the annual dry-matter production, Y (t ha^{-1} y^{-1}) may be expressed by

$$Y \approx 30/\{1+\exp(1.315-0.119T)\} \qquad 9.21$$

and

$$Y \approx 30\{1-\exp(-0.000664P)\} \qquad 9.22$$

where T (°C) is the mean annual temperature (−13 to 30 °C) and P (mm) is the mean annual precipitation (0 to 3500 mm). From the 149×10^6 km^2

Table 9.1. Total world production of some of the major crops and the mean yields in 1948/52 and 1967. Estimated percentage water content given in parentheses. After FAO Production Yearbook No. 22 (1968).

	Total crop (10^6 t)		Mean yield (t ha^{-1})	
	1948/52	1967	1948/52	1967
Wheat (15)	170.9	298.0	0.99	1.34
Rye (15)	37.0	31.6	0.96	1.39
Barley (15)	59.2	118.3	1.13	1.67
Oats (15)	61.7	50.7	1.14	1.63
Maize (15)	139.4	264.5	1.59	2.48
Millett and Sorghum (15)	48.3	90.6	0.51	0.79
Paddy rice (15)	167.3	275.9	1.63	2.14
Potatoes (80)	247.9	307.9	10.90	13.40
Sweet potatoes and yams (80)	69.9	135.3	6.80	8.50
Cotton (lint) (15)	7.5	10.4	0.24	0.33
Total pulses (15)	29.3	39.6	—	—
Total oilseeds (15)	51.3	98.0	—	—
Sugar				
centrifugal raw	32.3	66.3	—	—
non-centrifugal	5.5	11.1	—	—

Table 9.2. Some of the highest recorded yields and growth rates for several crops. Usually shoot weights only have been recorded; these values have been multiplied by 1.18 to give the entries under total dry matter. Marketable produce is expressed as dry matter. Values in brackets are estimates. Mainly from data collated by Stewart, G. A. (1970). *J. Aust. Inst. agric. Sci.* **36**, 85–101.

Species	Country	Length of growing season (d)	Crop growth rate g m^{-2} d^{-1}		Yield (t ha^{-1})	
			Average	Maximum	Total dry matter	Marketable produce
Napier grass	Kenya	365	24	50	104	88
Sugar cane	Hawaii	365	18	37	78	66
Sorghum	Northern Australia	360	13	—	52	46
Rye grass	Holland	186	12	20	26	22
Maize	Illinois, USA	132	23	—	[30]	12
	Iowa, USA	141	11	23	19	8
Rice	NSW, Australia	190	17	[40]	32	12
	Philippines	122	—	27	—	9
Potatoes	Holland	130	17	24	22	19
Sugar beet	California, USA	300	14	—	43	[29]
Wheat	Netherlands	[120]	16	—	20	6
Soybeans	Illinois, USA	120	16	—	10	3

Table 9.3. Crop yields in England and Wales, 1926–60. After H. W. Howard in Ivins & Milthorpe (1963).

Crop	Harvested component	Average yield of component (t ha⁻¹)			
		1926–30	1936–40	1946–50	1956–60
Wheat	Grain (*c.* 15 % water)	2.18	2.23	2.47	3.30
Barley	Grain (*c.* 15 % water)	2.06	2.06	2.31	3.01
Sugar beet	Roots (*c.* 75 % water)	20.4	22.4	25.4	32.8
Potatoes	Tubers (*c.* 80 % water)	16.0	17.4	17.4	19.8

continental surface and 510×10^6 km² oceanic surface, about 10^{11} and 0.5×10^{11} t dry matter are produced annually; of the continental production about 10 per cent is agricultural, produced from the 10 per cent of the land surface that is cultivated. (Errors of 50–70 per cent in these estimates are likely.) The margins of agriculture are set by a length of growing season which allows a production of about 1.5 t ha^{-1} or about 300–400 kg marketable produce ha^{-1}; within the cultivated area production is often restricted by transient water shortage, inadequate amounts of fertilizers, diseases and insect pests, and other factors. The actual production of the principal crops is shown in Table 9.1.

It is also of interest to examine the upper limit, considering the growing season only and the provision of adequate water and mineral nutrients (Table 9.2). These entries are for maximum achievements; growth rates and yields in the particular seasons, i.e. average crops of the area, rarely reach half these values and long-term averages from comparable regions may be less than one-quarter (Table 9.3). Calculations assuming a requirement of 10 light quanta per mole of carbohydrate formed, complete interception of light, and respiration rates equal to one-third photosynthesis indicate an absolute limit of 3.35 μg carbohydrate J^{-1} which with a maximum total radiation of about 350 W m^{-2} fixes the upper limit to about 100 g m^{-2} d^{-1}. Allowance for incomplete light interception and its distribution within a full canopy and decrease in efficiency with increase in light intensity, reduces this to about 50–70 g m^{-2} d^{-1}. Incomplete cover and, hence, light interception, and sub-optimal temperatures result in the average for the growing season being about one-half of the maximum growth rate. These data represent the highest yields that have been achieved in the different areas with existing varieties and present methods of agronomic management. Although no well-proven procedure for segregating the factors responsible for these limits yet exists (cf. Chapter 10), it seems likely that the highest (obtained with a near-optimal environment for the whole year) is only about three-quarters of that set by the supply of light and the current efficiency of the photosynthetic reactions.

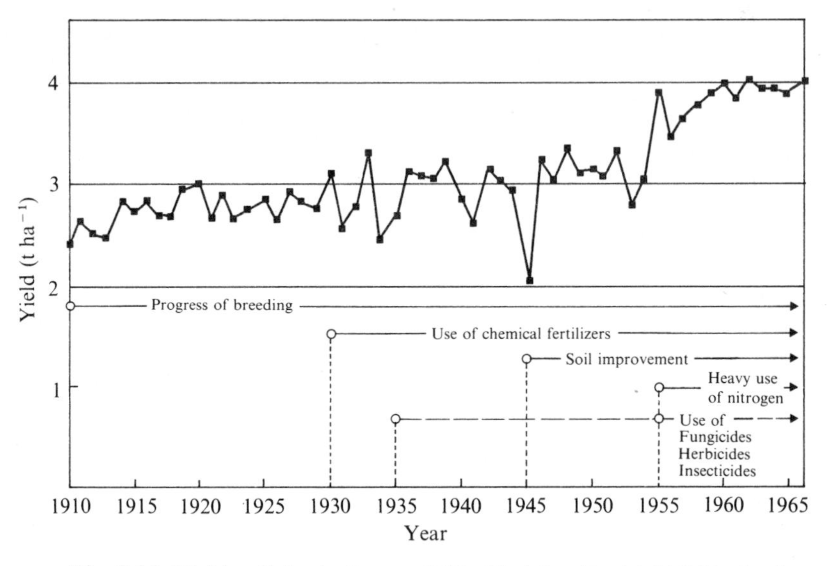

Fig. 9.16. Yields of rice in Japan, 1910–66. After Yoshiaki Ishizuka in Eastin *et al.* (1969).

The extent to which average production falls below the maxima may be seen from Table 9.1 and, more particularly, from two examples taken from highly intensive production systems. The first concerns rice production in Japan and shows the increase in yield during the past few decades and some of the responsible factors (Fig. 9.16). It may be noted that the average yield in 800 AD was about 1 t ha^{-1} and this increased to 1.9 t ha^{-1} in the early eighteenth century with the widespread use of irrigation systems. The average yield is now about half of that obtained on some of the experimental stations and less than one-third of the maximum attained. The second example shows changes in yield of some crops in England and Wales (Table 9.3). With wheat and barley (as with rice, maize and other cereals elsewhere) the yield has been substantially increased over the past two decades and is now about half of that obtainable from the existing varieties. The large increases in yield of grain have arisen from selecting varieties with short strong straw, low tillering capacity, ability to use high supplies of nitrogen without lodging, and differentiation of inflorescences over a long period, thereby resulting in about half the dry matter produced being diverted in grain. The higher yields of rice than wheat seem to reside in the differentiation of much larger inflorescences and the maintenance of photosynthesizing leaves until the time of harvest. The yield of sugar beet in England is less than that attainable in California (Table 9.2) because of the shorter growing season. Selection for high concentrations of sugar coupled with high yields of roots has not been particularly successful; where high root yield has been attained (as with the mangold varieties) the translocated sugar, which might otherwise have been stored, has been diverted to cell division and growth. Potato yields are also low. This arises from a number of causes, including the harvest of about 15 per cent of the total acreage at very low yields (but very high market prices). These yields are about half that of an average well-managed main-season crop and about one-quarter of the maximum attained. Nevertheless, for a long period, selection of new varieties was not particularly successful and until very recently about 70 per cent of the British acreage was planted with varieties introduced early in this century.

Finally, some comment should be made on the comparatively low yields obtained from indeterminate-growing species such as soybeans. The reasons for this appear to be associated with the lack of adequate flower and fruit set rather than with the ability of the plant as a whole to grow at an adequate rate. Fuller discussion of the issues involved with the attainment of high yields in this (and other) species is given in the companion volume edited by L. T. Evans and referred to in the Preface.

FURTHER READING

Donald, C. M. (1963). Competition among crop and pasture plants. *Adv. Agron.* **15**, 1–118.

Eastin, J. D. *et al.* (1969). *Physiological aspects of crop yield.* Amer. Soc. Agron., Madison, Wisconsin.

Eckardt, F. E. (ed.) (1968). *Functioning of terrestrial ecosystems at the primary production level.* UNESCO, Paris.

Loomis, R. S., Williams, W. A. & Hall, A. E. (1971). Agricultural productivity. *Ann. Rev. Pl. Physiol.* **22**, 431–68.

Milthorpe, F. L. (ed.) (1961). *Mechanisms in biological competition. Symp. Soc. exp. Biol.* **15**. Cambridge University Press.

Steward, F. C. (ed.) (1969). *Plant physiology*, Vol. 5A. Academic Press, New York and London.

Watson, D. J. (1952). The physiological basis of variation in yield. *Adv. Agron.* **4**, 101–45.

Yoshida, S. (1972). Physiological aspects of grain yield. *Ann. Rev. Pl. Physiol.* **23**, 437–64.

10

The Prediction of Responses

SOME GENERAL ISSUES

In the preceding chapters we have tried to summarize the main features of crop growth and the ways in which these interact with the environment. Those who have worked through the book should have emerged with some concept of the complex, dynamic interrelated nature of this system. Certain parts are understood in considerable detail and the general shape of the fabric and the way it fits together are known reasonably well. What is lacking is an adequate understanding of how all the components interact to determine the rate of change and the ultimate size of the whole; just what is the effect of a change in one part, say making leaves more upright, on the behaviour of the whole or on a component such as yield of grain. We have yet to face the task, stated in Chapter 1, of putting the relationships together 'in such a way that the magnitude in any stated set of circumstances can be predicted, and of then assessing the interplay between the different component processes so that the behaviour of the whole can be predicted'.

It will be clear that the approaches described in Chapter 9 are not adequate for this task – partly because the variables measured (such as growth rate) depend so much on interrelationships at a lower level of organization and, hence, are not readily predicted, and partly because of a lack of suitable methodology. Until recently, the only feasible approach for relating a measured variable, Z, such as net assimilation rate, say, to a complex of varying determining factors, X_1, X_2,..., X_n has been that of multiple regression. The established relationship may be regarded as the sum of a series of polynomials in the form

$$Z = a+b_1X_1+b_2X_1^2+\ldots+u_1X_n+u_2X_n^2 \qquad 10.1$$

each independent variable being established at the mean values of the others. The labour of computation has usually militated against extending each polynomial beyond a quadratic. Although useful in giving a crude indication of a relationship in poorly understood complex systems, multiple regressions have not been particularly successful in most situations (rarely accounting for more than 70–80 per cent of the variation) and are certainly not equal to our current need.

The advent of large-capacity high-speed computers has provided the technology required; all that is now necessary is to devise a suitable representation of the system in schematic form and with definable boundaries. Such can be provided by a simulation model using either analogue or digital techniques. An analogue model is one in which reality is represented by, usually, a comparable electronic circuit, the variables of the model and the real system being so scaled that quantitative answers are yielded. An example is outlined in Fig. 3.1. They can be very useful where the system is continuous and all functions are of a similar nature. Where a number of the steps are discontinuous, many empirical functions have to be used and, where changes depend on previous history as well as the current state of the system, a direct simulation using a digital machine is preferred. Such quantitative simulation models may be either 'stochastic' or 'deterministic'. Stochastic models describe events in statistical terms whereas deterministic models are essentially more definite and realistic, even though many of the relations used are purely empirical and are statistical in origin. A good example of the two different approaches to a single problem is that of describing electron orbits by either wave mechanics (stochastic) or Bohr's model of steady state orbits (deterministic).

The model must produce a satisfactory quantitative description of the behaviour of a particular crop in the variable environment in which it is grown and it should be capable of predicting the reactions of the whole and parts of the system over the range of conditions experienced. A likely consequence is that interrelationships or facets of the system hitherto unsuspected will be revealed. The model used must be verified and it should itself suggest useful experiments which can test and increase understanding of the system. This cycle of '—idea—model—prediction—verification—idea—' or '...non-verification—modified idea...' is an essential facet of modelling since no model can be better than the data or the concepts on which it is based. Although no model can be considered to be a final statement, or modelling an easy way of avoiding experimentation, an adequately verified model can substitute for a large number of costly and tedious routine field experiments.

Taking the levels of organization of a biological system as 'molecules–organelles–cells–tissues–organs–individuals–populations–communities', then it would seem technically difficult to simulate a system over more than two or three of these levels. That is, in our context, with the object of describing the behaviour of a population of similar individuals, we could not describe events below organ level. However, there is no restriction on simulating lower levels of organization separately; indeed, a sound model at a particular level – say, plant growth – should ideally rely on generalized statements from a number of much more detailed models – say, of ion uptake as influenced by root concentration, membrane properties, diffusion, release and fixation of ions, etc. Another technical restriction is that it is likely to be unprofitable – both in terms of accuracy of the end result as well as in cost of computer time – to work in time intervals shorter than 10^{-3} of the overall time. For example, in modelling growth of a crop over a growing season of 150 days, we would probably have to decide between using hourly steps or day and night periods. The latter should be perfectly adequate for most purposes. Such a model would incorporate a general daily function describing the water status, but this function could itself derive from detailed descriptions of the changes over a day, at say ten-minute intervals, in radiation interception and stomatal resistances in the different leaf layers, humidity profiles, water flow, soil water and root concentration in the different soil layers, etc.

Usually, a model of the type we are concerned with here takes the form of a successive series of integrations of the various relationships progressing in a step-wise manner. This type of model can be programmed in Fortran or an equivalent language. It must be stated, however, that this is not the only way of modelling and may not be always the most suitable. In many instances the use of a specialized simulation language such as CSMP is more convenient and involves less labour. These languages also allow operations to progress in parallel more readily than does Fortran and the use of various sophistications such as making the size of the time interval dependent on the rate of change.

DEVELOPMENT OF A MODEL OF CROP GROWTH

In the past few years an increasing number of attempts have been made to produce deterministic models of various aspects of plant growth. There are now fairly comprehensive models of the water balance in the soil, the absorption and evaporation of the water by plants, and of light interception and photosynthesis. Many of these are described or cited in Šetlík (1969), a volume which will repay close study. It is of interest to note that in that volume, dealing with the measurement of photosynthetic productivity, there are no papers on mineral nutrients and only two attempts to deal with the sequential changes during growth. This reflects the excessive attention given to light interception and photosynthesis at the expense of integration of the whole and the role of other supply factors of equal importance. This neglect is now being remedied, and the current literature includes an increasing number of papers dealing with the simulation of crops as influenced by the whole of their environment.

It is not apposite in this context to present the components of a model of crop growth in full detail; indeed, if such were done it would possibly turn out to be a summary, in symbolic language, of most of what has been written before. However, it does seem worth while to examine some of the issues

Table 10.1. Minimum amount of information for the production of a model of crop growth.

(*a*) *Aerial environment*

1. Mean day and night temperatures.
2. Total amount of visible radiation in each photoperiod.
3. Total net radiation in each photo- and dark-period.
4. Length of photoperiod.
5. Profile of visible radiation through crop canopy.
6. Profile of net radiation through crop canopy.
7. Profile of dry bulb temperature through crop canopy.
8. Profile of water vapour content of air through crop canopy.
9. Daily wind run.
10. Rainfall.

(*b*) *Soil environment*

1. Amounts of water in soil layers at the start of simulation.
2. Soil water content at wilting point ($\tau = -1500$ J kg^{-1}) and field capacity ($\tau = -10$ J kg^{-1}) in each of the soil layers.
3. Amounts of available N and P in each soil layer at the start of simulation.
4. Temperature profile through soil.
5. Rates of fixation and release of N and P in soil.

(*c*) *Crop characteristics*

1. Dry weights of meristems in seed and of stem, leaf and root primordia at some predetermined stage of growth considered to be time of initiation.
2. Number of seeds per unit area.
3. Relative concentrations of photosynthate, N and P in meristematic tissues and the change in these concentrations as the tissues age.
4. Values of constants in logistic equations used to describe growth and variations in growth of organ with temperature.
5. The weight of leaf primordia in dicotyledonous plants at unfolding from apical bud or the fractional size of monocotyledonous leaves at emergence from the next older leaf sheath.
6. Size of root members at branching and rate of production of branches.
7. Proportionality factors relating leaf weight to area, internode weight to length, and root length to weight.
8. Amounts of reserve materials required in pool for start of branch growth.
9. Requirements for onset of flowering.

involved in so integrating the approaches and material summarized in the preceding chapters, so as to arrive at an effective simulation of crop growth.

At this rudimentary stage in the art of simulation, it is realistic to think of different models for each crop species. These can then take into account certain of the basic genetic features, such as tillering, grain or tuber production, periodic defoliation (e.g. lucerne), etc., and remove a number of complexities which would be required in the simulation of a 'general' plant. Of course, some parts of all models – such as those dealing with water and photosynthesis – will be similar, although the values for the parameters included in the various functions may well be different. Having selected our species and the level of organization at which we are working, we then have to decide the degree of detail required. This will be decided primarily by the amount and accuracy of the data available and the degree of precision required in the end result. For example, the model required to estimate the yield of grain of wheat from a region with widely variable rainfall and a range of soil and management procedures may rely on water only and will be much shorter and more general (and less precise) than one designed to describe how the whole system is integrated and for which a reasonable array of basic inputs is available.

To narrow our terms of reference, we will now proceed to consider the outlines of a model describing the growth and yield of wheat in an area with a growing season of about 150 days and for which reasonable standard environmental inputs are available. Growth proceeds in steps at the end of each light and dark period. It is convenient to consider the *ontogeny* of the crop as conditioned by the supply of *water*, *mineral nutrients* and *photosynthate*, these being influenced by the current *environment*, and to divide the model into five sections or sub-routines dealing with each. Let us look at each in turn:

Environment sub-routine

This takes basic metereological and soil inputs and from known functions produces the quantities listed under (*a*) and (*b*) in Table 10.1.

Ontogeny sub-routine

The main components required are given in Table 10.1(*c*). A wheat seed has a testa surrounding an endosperm and an embryo consisting of a stem apex, three leaf primordia and three seminal roots. The dry weight of each part is entered for seed of known mean weight and the initial growth consists of the transfer of dry matter from the endosperm to the embryo.

The growth of any determinate organ is described by a simple logistic equation (cf. Equation 7.2)

$$W = A\{/1+b\exp(-kt)\} \qquad 10.2$$

where A is the maximum size attained by the organ under optimal conditions and the rate at which the organ grows is controlled by the constants b and k. Any reduction in k leads to a decrease in A; k is a function of temperature and varies with organ and position (cf. Equation 7.3). We base our allocation of dry matter, and hence growth, on the balance between a series of demands derived from Equation 10.2 and the supply of the necessary materials simulated in the supply routines.

The potential demand is determined by k/k_{max}, where k_{max} is the value of k for the organ when all environmental factors are optimal, this being the same for all organs. The use of appropriate values of k and A in Equation 10.2 gives the potential demand for dry matter by the organ at the temperature and water status pertaining. This demand can be converted into demands for assimilates, nitrogen and phosphorus. The conversion assumes that in meristematic regions there is a constant relationship between dry matter and the major mineral nutrients (about 10 per cent nitrogen and 1.3 per cent phosphorus). This proportionality changes as the cells extend, falling linearly to 3 per cent and 0.4 per cent at the time the leaf reaches full expansion. (N.B. These are 'demand' factors and do not necessarily reflect the actual concentrations in an organ at any time, the latter being adjusted later.)

These demands are compared with the supplies of assimilates, nitrogen and phosphorus provided by the other routines. If there is a deficiency in any item it can be withdrawn from a pool of reserves established when supplies were more plentiful. If the pool has been depleted the allocation of materials in short supply is made in the order: for assimilates – stem apex, leaves, stems and roots; and for nitrogen and phosphorus – stem apex, roots, stems and leaves. The limiting supply is then used to recalculate a new A which, in turn, will generate new demands for the materials in ample supply. Any surpluses are transferred to the pool and can be called on by other organs or in future cycles of the model.

The stem apex is assumed to grow to a critical size (0.0006 mg); it then divides to produce a new leaf (0.0002 mg), an internode (0.0001 mg) and a branch (0.0001 mg). The remaining 0.0002 mg is the new stem apex which continues to grow as previously. The increase in leaf size depends on materials from other leaves until it reaches a certain size when it becomes photosynthetic. In wheat emergence begins when leaf weight is $0.1A$ and all the leaf becomes photosynthetic when it is A. Leaf area is calculated as a function of leaf weight and the water and mineral nutrient status.

Stem growth can be considered as the elongation of a series of internodes up to a certain weight (in wheat $0.33A$); after this time any increase in weight produces thickening but no further elongation. By accumulating the internode lengths and leaf areas it is possible to determine the leaf area profile for the supply routines.

Tillers start growing when the amounts of photosynthate or nitrogen in the pool reach a critical level. Once growth starts, each branch behaves in the same way as the main stem, but drawing on the pool for its supplies until it has produced roots from its older internodes and photosynthetic leaves.

The roots are 'grown' in the same way as the shoots, following the relationships outlined on p. 134. Flowering takes place according to a routine which depends on photoperiod, temperature or time, according to variety.

Senescence of a leaf is programmed by assuming that when it reaches a maximum area it can no longer import materials from elsewhere and that

r'_m starts to increase, leading (together with shading by upper leaves) to its being unable to supply its own requirements for respiration. Its nitrogen and phosphorus are also exported if these are in short supply. Tiller death in cereals and grasses is programmed in a similar fashion. Inflorescence and grain growth are simulated in a similar way based on features outlined in Chapter 8.

Water sub-routine

This routine has to provide estimates of actual and potential evaporation by the crop, the relative leaf water content in the different layers of the canopy and the water content of the soil layers. In order to provide this we have to know the net radiation absorbed by the crop and soil, the resistances to water vapour transfer of the leaf layers of the crop and the boundary layer, and the coefficient for turbulent transfer. The water that is evaporated comes from different layers in the soil and the extraction pattern is a function of the length of roots and the soil water potential in each layer (cf. Chapter 3).

Photosynthesis sub-routine

The rate of gross photosynthesis is calculated for each leaf as a function of the irradiance, using a relationship such as Equation 5.8 or a simpler alternative. The effect of the water status of the plant is exerted through changes in r'_s and those of mineral nutrition and ontogeny through r'_m. The total respiration can be obtained by using the relationships described in Chapter 5, and is subtracted from the total gross photosynthesis to provide the supply of photosynthates which can be allocated according to the demand functions generated in the ontogeny sub-routine, any surplus being stored in the reserve pool.

Mineral nutrient sub-routine

The amounts of available nitrogen and phosphorus at the start of the growing season are entered and the total uptake in each time interval computed as a function of the concentrations of available ions and the root lengths in the various soil zones. The nutrients not used in the roots or allocated to the other organs according to the 'demands' of these organs are assigned to the reserve pool.

The phosphate moves to the roots by diffusion, and it can be assumed that the root absorbs all the available phosphate in a volume of soil of equal radius to the root hair zone within five days of the root entering that region of soil. It is necessary, therefore, to keep a running total of the root lengths generated in the soil zones and the ages of these roots. For reasonably sparse root systems it is probably sufficient to assume that the amount of available phosphate remains constant throughout the growing season. For denser root systems and situations where fertilizer is applied during the growing season, changes in the amount of available phosphate have to be simulated.

A comprehensive model of the interconversions of nitrogen in the soil would be extremely complex (cf. Fig. 2.2), but in many instances it is sufficient to make the amount of available nitrogen a function of the soil water content and the temperature.

CONCLUSION

This now brings us to the end of our story. We have tried to present a coherent, albeit terse and abbreviated, account, and are conscious of many facets for which the reader will need to search more deeply elsewhere. Most of the space has been devoted to looking at the more detailed parts of the whole system because future progress depends on understanding it fully and being able to integrate it adequately. The cohesion of the whole and its several parts into rigorously interrelated quantitative descriptions is possibly now the most important immediate issue and is receiving an accelerating degree of attention. This will in turn lead to further reductionist (i.e. Cartesian) experimentation and increased understanding; it also opens up completely new approaches in field experimentation and plant selection, and will almost certainly revolutionize many of the approaches and practices that have become traditional. Those engaged in this work will, therefore, find many new and exciting challenges.

FURTHER READING

Eastin, J. D. *et al.* (1969). *Physiological aspects of crop yield.* Amer. Soc. Agron., Madison, Wisconsin.

Eckardt, F. E. (ed.) (1968). *Functioning of terrestrial ecosystems at the primary production level.* UNESCO, Paris.

Lemon, E., Stewart, D. W. & Shawcroft, R. W. (1971). The sun's work in a cornfield. *Science, N.Y.* **174**, 371–8.

Loomis, R. S., Williams, W. A. & Hall, A. E. (1971). Agricultural productivity. *Ann. Rev. Pl. Physiol.* **22**, 431–68.

Šetlík, I. (ed.) (1969). *Prediction and measurement of photosynthetic productivity.* Proc. IBP/PP Tech. Meeting, Trebon, 1969. PUDOC, Wageningen.

List of Symbols and Units Used

Mathematical Symbols

log	Logarithm to base 10
ln	Logarithm to base e
exp x	Exponential of $x = e^x$ where e is exponential index (= 2.7183)
π	Ratio of circumference of circle to its diameter (= 3.1416)
$\Sigma(x)$	Sum of x
Δx	Finite increase of x
δx or dx	Variation of x
$[x]$	Concentration of x

Units

SI units are used wherever possible. The most commonly used units are given in the list.

Prefixes used with units

p, 10^{-12}; n, 10^{-9}; μ, 10^{-6}; m, 10^{-3}; c, 10^{-2}; d, 10^{-1}; da, 10; h, 10^2; k, 10^3

Abbreviations for units

d	Day
E	Einstein (= 6.02×10^{23} photons)
fc	Foot candle. A unit of illumination, now rarely used (= 10.76 lux), in which the irradiance depends on the spectral characteristics of the light source.
g	Gram
h	Hour
ha	Hectare (10^4 m^2)
J	Joule (= 0.239 calorie)
m	Metre
	Minute
s	Second
t	Tonne (10^3 kg)
w	Week
W	Watt
y	Year

Other symbols

A	Area (m^2)
A_x	Cross-sectional area of xylem
B	Position of leaf on stem axis numbered from base
D	Diameter (m)
E	Energy
	Quantity of exchangeable ions in soil (mg (g of soil)$^{-1}$)
	Net assimilation rate (g dm^{-2} w^{-1})
E_p	Potential net assimilation rate
I	Irradiance (W m^{-2}). In photosynthetic context refers to 300–700 nm radiation band
I_a	Absorbed radiation
I(0)	Irradiance at crop surface
I_d	Diffuse visible radiation
I_0	Direct visible radiation
K	Dimensionless parameter describing effect of leaf geometry and solor elevation on attenuation of radiation, such that $K = k(1-\gamma)$
K_d	Transmission coefficient for diffuse radiation in crops
K_s, K_m	Measure of affinity of uptake site; the concentration of the ion K_s, or carbon dioxide, K_m, when the rate of uptake is half maximum
L	Length (m)
M	Mass (kg)
M_a	Mass of air in soil
M_s	Mass of dry soil
M_w	Mass of water in soil
N	Hours of daylight
	Number
P	Pressure potential due to head of water in soil (J kg^{-1})
	Turgor pressure (J kg^{-1})
	Rainfall (mm)
R	Gas constant (8.315 J °K^{-1} mole^{-1})
	Radius (m)
$\bar{R}$	Mean radius of root members
S	Fraction of soil volume filled with air
S_G	Dry matter used in growth
S_M	Dry matter used in maintenance respiration
S_R	Dry matter lost by plant during respiration
T	Temperature (°C unless specified as °K)
T_s	Surface temperature
V	Volume
V_a	Volume of air in soil
V_s	Volume of solids in soil

W Mass (g) after drying to constant weight
W_F Fresh weight (g)
Y Yield ($kg\ m^{-2}$)
X, Z Any entity in general

A	Asymptote in growth curves, e.g. notional maximum yield in fertilizer response experiments
	A distance from the root surface $\approx 1/(\pi R_v)^{0.5}$
C	Concentration, including a gas in the atmosphere or an ion in the bulk soil solution ($g\ g^{-1}$; $g\ cm^{-3}$)
C_a, C_{chl}, C_w	Concentration of carbon dioxide in atmosphere, chloroplast and mesophyll cell wall of leaf, respectively
C_N	Concentration of total organic nitrogen in soil
C_r	Concentration of ion at root surface
C_s	Molal concentration (moles kg^{-1})
D	Diffusion coefficient
E	Flux density of water vapour ($kg\ m^{-2}\ s^{-1}$ or $mm\ d^{-1}$)
E_p	Potential transpiration
F	Flux density of an entity ($g\ m^{-2}\ s^{-1}$ or $m^3\ m^{-2}\ s^{-1}$)
F_d	Diffusive flux of ions in soil solution
F_f	Convective flux of ions in soil solution
H	Flux density of sensible heat ($J\ m^{-2}\ s^{-1}$)
K	Transfer coefficient ($m^2\ s^{-1}$)
K_H	Transfer coefficient of sensible heat
K_M	Transfer coefficient of momentum
K_P	Transfer coefficient of carbon dioxide
K_W	Transfer coefficient of water vapour
L	Leaf area (cm^2); leaf area per unit area of soil ($cm^2\ cm^{-2}$, sometimes called leaf area index)
L_v	Leaf area density ($cm^2\ cm^{-3}$ of canopy)
M	Momentum ($kg\ m^{-1}\ s^{-2}$)
P	Photosynthesis, flux density of carbon dioxide ($mg\ CO_2\ dm^{-2}\ h^{-1}$)
P_G	Gross photosynthesis
P_N	Net photosynthesis
Q	Radiant flux density ($W\ m^{-2}$)
Q_A	Incoming short-wave radiation at top of atmosphere
Q_L	Long-wave radiation
Q_N	Net radiation ($= Q_A + Q_L$)
$Q_N(0)$	Net radiation above crop
$Q_N(S)$	Net radiation at soil surface
Q_0	Radiation of defined spectral composition at crop surface
Q_S	Short-wave radiation (0.3–3 μm)
Q_V	Visible radiation (300–700 nm)

R	Respiration (mg CO_2 dm^2 h^{-1})
	Relative growth rate (d^{-1})
R_G	Growth respiration
R_M	Maintenance respiration
R_V	Root concentration (cm root (cm^3 of soil)$^{-1}$)
S	Concentration of sorbed ions
X	Concentration of an ion in the plant
a, b	General constants
b	Maximum efficiency of conversion of radiation in photosynthesis (m^2 W^{-1})
c	Fraction of sky covered by cloud
c_p	Thermal capacity of moist air at constant pressure (1 kJ kg^{-1} °C^{-1})
d	Crop density (plants m^{-2})
e	Vapour pressure of water in air (mbar) (1 mbar = 10^2 newton m^{-2})
e'	Saturated vapour pressure of water in air (at a specified temperature)
f	Constant in Penman equation (dimensionless)
g	Acceleration due to gravity
h	Conductivity of root cortex to water (kg cm^2 s^{-1} J^{-1})
	Planck's constant (6.625×10^{-27} erg s)
j	Buffering capacity of soil (equiv g^{-1})
k	Attenuation coefficient of light through a crop canopy (dimensionless)
	Capillary conductivity of soil (kg cm^2 s^{-1} J^{-1})
	Constant relating spectral composition of radiation and K, concerned with crop geometry
k_X	Conductivity of xylem for flow of water
m	Maintenance coefficient (in respiration)
n	Hours of bright sunshine received per day
	Fraction of CO_2 released in respiration that goes into intercellular spaces
p	Per cent germination
q	Specific humidity (g water vapour (g of moist air)$^{-1}$)
	Rate of water uptake per unit length of root
q'	Specific humidity at saturation
q'_s	Specific humidity at saturation at the evaporating surface of the leaf
r	Resistance to water vapour transfer (s cm^{-1})
r'	Resistance to transfer of carbon dioxide
r_L	Resistance of lower surface

r_U	Resistance of upper surface
r_a	Boundary layer resistance
r_c	Cuticular resistance
r_h	Resistance to heat transfer
r_i	Resistance of sub-stomatal cavities
r_l	Leaf resistance
r_m	Mesophyll resistance
r_s	Stomatal resistance
r_w	Resistance of mesophyll cell walls
r'_x	Excitation resistance
t	Time
u	Wind velocity (m s^{-1})
u_*	Friction velocity
v	Velocity (m s^{-1})
	Volume flux ($m^3\ m^{-2}\ s^{-1} = m\ s^{-1}$)
v_s	Volume flux of water at soil surface
z	Vertical dimension (m)
z_0	A roughness coefficient relating to the displacement of the wind profile above the surface (approximately 0.1 of height of protuberances above surface)
Γ	Intercellular concentration of carbon dioxide
Φ	Hydraulic potential of soil water (J kg^{-1})
Ω	Efficiency of production of new growth
α	Reflection coefficient
	Efficiency factor for ion uptake by roots; $\bar{\alpha}$, mean value over whole root surface
β	Effective emissivity of sky for long-wave radiation
γ	Psychometric constant (0.66 mbar $°C^{-1}$)
	Transmission coefficient of radiation through leaves
ϵ	Ratio of increase in latent heat content per unit increase in sensible heat
ζ	Relative leaf water content (per cent)
η	Viscosity of water (10^{-2} g $cm^{-1}\ s^{-1}$ at 20 °C)
θ	Water content of soils or plant tissues (g (g dry wt)$^{-1}$)
λ	Latent heat of evaporation of water (2453 J g^{-1} at 20 °C)
	Wave-length of radiation
ν	Frequency of radiation (s^{-1})
ξ	Factor used to modify the diffusion equation to allow for uptake in a radial system
π	Osmotic potential (J kg^{-1})
ρ	Density (g cm^{-3})

ρ_a	Density of air (1.2 mg cm^{-3})
ρ_s	Bulk density of soil
ρ_w	Density of water
σ	Surface tension of water
	(72.8 dyn cm^{-1} = 72.8×10^2 N m^{-1})
	Stefan–Boltzmann constant (5.67×10^{-8} W m^{-2})
τ	Matric potential (J kg^{-1})
τ_s	Matric potential in soil remote from root
τ_0	Matric potential at root surface
ϕ	Net flux of radiation to leaf
χ	Stability length referring to turbulent transfer
ψ	Water potential (J kg^{-1})
ψ_l	Water potential of leaves
ψ_r	Water potential of roots
ψ_s	Water potential of soil
ψ_x	Water potential of xylem
ω	Overall efficiency of assimilate utilization

Index